The Best of Borneo Travel

'Nepenthes Edwardsiana', from Spenser St. John, *Life in the Forests of the Far East*, Smith, Elder and Co., London, 1862.

The Best of Borneo Travel

Compiled and Introduced by
VICTOR T. KING

SINGAPORE
OXFORD UNIVERSITY PRESS
OXFORD NEW YORK
1992

Oxford University Press

Oxford New York Toronto
Delhi Bombay Calcutta Madras Karachi
Kuala Lumpur Singapore Hong Kong Tokyo
Nairobi Dar es Salaam Cape Town
Melbourne Auckland Madrid
and associated companies in
Berlin Ibadan

Oxford is a trade mark of Oxford University Press

ISBN 0 19 588603 8

British Library Cataloging in Publication Data

A catalogue record for this book is
available from the British Library

Library of Congress Cataloging-in-Publication Data

The Best of Borneo travel/compiled and introduced by Victor T. King.
p. cm.—(Oxford in Asia paperbacks)
ISBN 0-19-588603-8:
1. Borneo — Description and travel.
I. King, Victor T.
II. Series.
DS646. 312. B47   1993
915.98'3 — dc 20
92 –26433
CIP

Typeset by Indah Photosetting Centre Sdn Bhd., Malaysia
Printed by Peter Chong Printers Sdn. Bhd., Malaysia
Published by Oxford University Press Pte. Ltd.,
Unit 221, Ubi Avenue 4, Singapore 1440

Preface and Acknowledgements

IT is necessary in an anthology with the word *best* in the title to say something about the process and criteria of selecting appropriate extracts. *The Best of Borneo Travel* is a grand title and a potentially misleading one. This collection is very much a personal one. Another compiler would, I feel sure, have chosen a rather different set of readings. Nevertheless, I am certain that among enthusiasts and experts there would be some measure of agreement about what constitutes the best in travel writing on Borneo. For example, I cannot imagine an anthology of this kind without pieces from the works of Charles Hose, Odoardo Beccari, Alfred Russel Wallace, Carl Lumholtz, Hendrik Tillema, Malcolm MacDonald, Tom Harrisson, Redmond O'Hanlon, and Eric Hansen.

I think it useful to explain why I have chosen what I have and to outline the demands and constraints on that choice. One must also bear in mind that selection was a two-stage process, first between the large number of books available on Borneo travel, exploration, and adventure, and second in extracting suitable passages from a book once chosen. As the reader might expect, there were immediate technical, financial, and commercial considerations. I have had to be especially selective because the publisher set its requirement for the length of the book; obviously costs and price have to guide us in our search for the best. What is more, each extract had to be, as far as possible, a substantial piece, which would stand on its own and make some sort of sense in its own terms. Thus, I attempted to find consolidated and defined sections of a book. In a number of cases a convenient and suitably sized chapter presented itself and made my task easier than it would otherwise have been. In other cases surgical operations were needed to cut and stitch together different sections of a book, or lift a short piece from a long chapter, so that the overall extract covered those aspects of the given author's work to which I wanted to draw attention, and illustrated sufficiently well the writer's concerns and style while retaining coherence and succinctness. To assist the reader, each extract is prefaced by a short introduction to the author and the context of the events, journey, or adventure described.

Given the considerations above, it was really only possible to contemplate 18–20 contributions at the most. There was yet another important concern. To keep to a financially manageable level of permissions fees for those books still in copyright, there was an obvious restriction on the number of recent travel books which could be included. Partly for this reason, the anthology comprises a majority of early texts, though, in my view, many of these happily do constitute the *best* examples available of Borneo travel writing.

Another factor influencing my choice was to try to cover what I would call the characteristic features of Borneo and its peoples, which have become lodged in the popular imagination of Western audiences. Perhaps some might say that this is an overly ethnocentric exercise, but I think it important in an anthology of this kind to convey what it was that was seen to 'define' Borneo for Europeans. Interestingly, it is still many of those features emphasized by early European travellers and explorers which are used currently to promote Borneo as a tourist destination and as a place worth seeing. These remain what the modern generation of travellers, tourists, and budding explorers want and, perhaps, even expect to see in Borneo. I have therefore selected items which describe the different environments of the island: the rain forest and shore; the fauna and flora, among them the orang-utan, Rafflesia, pitcher plants, and durian; rapids and river journeys; and upland ranges, particularly the highest mountain in Borneo and, indeed, in the whole of South-East Asia, Kinabalu, in the Malaysian state of Sabah. Of course, there also had to be accounts of interior pagan Dayak[1] longhouses,

---

[1]The general term 'Dayak' is commonly used to refer to the non-Muslim, non-Malay indigenes of Borneo. Among others, the ethnic groups which comprise the category 'Dayak' include the Kenyahs, Kayans, Kelabits, Ot Danums, Ngajus, Ibans (Sea Dayaks), and Bidayuhs (Land Dayaks). Hunter-gatherers in Borneo are sometimes referred to as 'Dayaks', or alternatively, 'Punans'. Some groups in Sarawak are called Penan.

Muslim-Malay coastal dwellings on stilts, head-hunting, and native customs, costume, and ceremony.

I also wanted to include contributions from different parts of Borneo, especially to ensure that some descriptions of Indonesian Kalimantan were included. This is difficult in an English language publication because many of the best writings on travel and exploration in the former Dutch territories of Borneo are only available in Dutch and German. Those that immediately spring to mind are the journeys of C. A. L. M. Schwaner through South, Central, and West Borneo in the 1840s, A. W. Nieuwenhuis's expeditions through Central Borneo in the 1890s, as well as Karl Helbig's crossing of the island from west to east in the 1930s. The choice for the vast territories of Kalimantan is therefore relatively limited, but I have managed to include extracts in English from Beeckman, Bock, Lumholtz, Tillema, Hansen, and Hanbury-Tenison. Still, the reader will see that a lot has been provided on the Malaysian state of Sarawak, partly because much has been written on it, given its early accessibility and openness to European exploration under the personal rule of Sir James and Sir Charles Brooke. There are only titbits, I am afraid, on Brunei and Sabah.

Finally, I felt that the anthology needed a spread of travel writings from different periods of time and from different kinds of European contact with Borneo. I have chosen two extracts which illustrate early visits to two of the most important Muslim sultanates and trading centres—Brunei and Banjarmasin—in the period of early European mercantilism. There is also ample illustration of journeys of exploration and adventure during the initial stages of the establishment of European administration and territorial control. Then, there are travels during the late colonial period when Europeans had a firm political hold on the country and, through military expeditions and pacification, had stamped out head-hunting, piracy, and warfare. Lastly, there are the most recent travel accounts from the period of national independence of the formerly British territories of Sarawak and North Borneo (Sabah) incorporated into the Federation of Malaysia in 1963, and the previously extensive Dutch domains in the south, brought into the Republic of Indonesia as Kalimantan, following independence in late 1949.

In conclusion, it might be useful to mention briefly some of the books which I considered including in the anthology, but, for the reasons which I have now fully explained, fell by the wayside. I do not wish to suggest that these are necessarily any less worthy of our attention than those eventually included. They were serious contenders and comprised Hugh Low, *Sarawak, its Inhabitants and Productions* (Richard Bentley, London, 1848); Ida Pfeiffer, *A Lady's Second Journey Round the World* (Longman, Brown, Green and Longmans, London, 1855); William T. Hornaday, *Two Years in the Jungle: The Experiences of a Hunter and Naturalist in India, Ceylon, The Malay Peninsula and Borneo* (Kegan, Paul, Trench and Co., London, 1885); John Whitehead, *The Exploration of Mount Kinabalu, North Borneo* (Gurney and Jackson, London, 1893); William H. Furness, *The Home-Life of Borneo Head-Hunters: its Festivals and Folklore* (Lippincott, Philadelphia, 1902); Robert W. Shelford, *A Naturalist in Borneo* (T. Fisher Unwin, London, 1916); Guy Arnold, *Longhouse and Jungle. An Expedition to Sarawak* (Chatto and Windus, London, 1959); Guy Piazzini, *The Children of Lilith* (Hodder and Stoughton, London, 1960); Jørgen Bisch, *Ulu. The World's End* (George Allen and Unwin, London, 1961); Mora Dickson, *A Season in Sarawak* (Dennis Dobson, London, 1962); Barbara Harrisson, *Orang-Utan* (Collins, London, 1962); Jean-Yves Domalain, *Panjamon* (Rupert Hart-Davis, 1972; and Panther Books, St. Albans, 1974); Wyn Sargent, *My Life with the Headhunters* (Arthur Barker Ltd. 1974; and Panther Books, St. Albans, 1976); James Barclay, *A Stroll Through Borneo* (Hodder and Stoughton, London, 1980); and David Macdonald, *Expedition to Borneo* (J. M. Dent, London, 1982).

For permission to reprint the extracts which I did include I am grateful to Oxford University Press, Kuala Lumpur/ Singapore, for facilitating publication by making the following available from its reprint series: Tom Harrisson, *Borneo Jungle*; Charles Hose, *The Field-Book of a Jungle-Wallah*; Odoardo Beccari, *Wanderings in the Great Forests of Borneo*; Charles Brooke, *Ten Years in Sarawak*; Spenser St. John, *Life in the Forests of the Far East*; Carl Bock, *The Head-Hunters of Borneo*; and Carl Lumholtz, *Through Central Borneo*; and for granting permission to use an extract from Hendrik Tillema, *A Journey among the Peoples of Central Borneo in Word and Picture* (1989, Oxford University Press, Singapore). I wish also to acknowledge the Council of the Hakluyt Society in reprinting an extract from *The First Voyage Round the World by Magellan*. Finally, for permission to reprint extracts from copyright material, I have to acknowledge, with thanks, Jonathan Cape and Random Century Group for Malcolm MacDonald's *Borneo People* (1956, pp. 101–13, 279–88); Hutchinson and Random Century Group for Tom Harrisson's *World Within. A Borneo Story* (1959, pp. 3–4, 22–5, 114–21; also available in Oxford's reprint series); the estate of the late Marika Hanbury-Tenison, A. M. Heath and Co. Ltd., Author's Agents, and Hutchinson for Marika Hanbury-Tenison's *A Slice of Spice. Travels to the Indonesian Islands* (1974, pp. 83–102); Peters Fraser and Dunlop Ltd. for Redmond O'Hanlon's *Into the Heart of Borneo* (Salamander/Penguin, 1984, pp. 75–89); John Murray (Publishers) Ltd. for Andro Linklater's *Wild People* (1990, pp. 99–110); and Hutchinson and Random Century Group for Eric Hansen's *Stranger in the Forest. On Foot across Borneo* (1988, pp. 153–74).

In writing the 'Introduction' I am grateful for the insights provided by Dr Graham Saunders in his paper, 'Early Travellers in Borneo', presented at a conference on 'Tourist Development in South-East Asia', under the auspices of the Association of South-East Asian Studies in the United Kingdom, held in Hull in late March 1991. I am also conscious of my debt to Dr Trevor Millum, who is currently working on the subject of travel writing on Borneo for a Master's thesis at the Open University, Milton Keynes. Dr Millum kindly showed me some of his draft chapters and invited my advice and comments.

I should also like to record my gratitude to the Oxford University Press for earlier giving me the occasion to undertake work on two of the travellers featured in this anthology: Hendrik Tillema and Carl Lumholtz.

Finally, in compiling the anthology, I have kept editorial changes to a minimum. However, in some cases I have omitted certain references, technical details, footnotes, and tables from the extracts, where these interrupt the flow of the narrative. The original prose of the various authors has been preserved in the interest of authenticity. No attempt has been made to edit the passages for consistency; the original punctuation and spelling have also been retained. Where ethnic terms, place names, and so on require transcription or clarification, this is indicated in my footnotes to the particular extracts. Appropriate illustrative material has been included, but this has not necessarily been taken from the books featured in this anthology. I have also compiled a map of Borneo so that the journeys described can be more easily traced.

Centre for South-East Asian Studies   VICTOR T. KING
University of Hull
March 1992

Contents

Introduction:
Travel and Travellers in Borneo

VICTOR T. KING

IN the European imagination the island of Borneo, the third largest island in the world, situated in the geographical centre of the Malay–Indonesian archipelago, has commonly been associated with mystery, danger, and excitement. These images were created a long time ago by European travellers, explorers, adventurers, and administrators, and, by the early part of the twentieth century, the popular conceptions of Borneo were already firmly in place.

This anthology of the *The Best of Borneo Travel* has been compiled to illustrate the preoccupations of Europeans, and the natural and cultural characteristics of Borneo which have caught their attention and which they have decided to record on their journeys. But what do I mean by travel and, more especially, writing about travel? Without getting into a detailed discussion of the literary and technical qualities of travel writing, and whether it can be defined as a particular genre, I adopted some simple rules of thumb during my task of compilation.

At its most basic, of course, travel writing involves a factual account of a specific journey, a passage from one place to another. It should also comprise an attempt to stimulate the imagination, emotions, and sensations so that, in some way, the reader experiences the events, has some sense of the place and the people described, and imaginatively participates in the adventure. In this regard, we are dealing with a personal record, in which the reader should feel involved. Given this requirement, we would not expect to include academic texts on the natural history or culture of Borneo, although some writings about travel do contain weighty materials of scholarly concern and interest. The works of Hose, Beccari, and Wallace, for example, precisely illustrate a combination of the personal journey and the results of scientific observation. In addition, the writer, in recounting a movement from place to place, and, often in the Bornean case, a passage into the unknown or unfamiliar, commonly does so in a narrative form.

I have to say, at this point, that in order to illustrate various aspects of the environment, peoples, and cultures of Borneo, I have not been able to provide a sense of passage or motion in all the pieces included. In some cases, the extract describes a sojourn in a place and what happened there, or explains or discusses an item of nature or a cultural trait, although usually in the broader context of travel. Furthermore, not every author in this anthology is writing as someone involved in a specific expedition, voyage, journey, or tour. Many are, but Charles Hose, Spenser St. John, and Charles Brooke are describing their experiences and observations not as individuals involved in a particular journey with a definite beginning and end, but rather as long-term residents of the country, living and working there. Yet, even in these cases, these representatives of European government had to do much travelling, and I have been able to choose pieces from their writing which involve 'excursions': Hose sailing along the Bornean coast, St. John climbing Mount Kinabalu, and Brooke engaged in a show of force against the Saribas Dayaks.

Having set down what I consider to be writings about travel, it remains to outline briefly the variations in the journeys included in this anthology. Individuals may travel for different reasons, although commonly there is a desire to see and experience new things. People may travel alone, or with one or few companions, or they may be part of a large expedition. Some early travellers have gone into areas previously unknown to Europeans; others have moved along relatively well-known paths. Certain of our authors reveal rather more of themselves and their opinions. This is especially so of the recent travel writers like Hanbury-Tenison and Hansen. Others are there, but as distant, dispassionate observers and reporters such as Lumholtz. Earlier travel writing, because it tended towards exploration and scientific discovery, was more likely to present seemingly objective and sober accounts of the journey. However, there are exceptions. Boyle's and Bock's accounts, although written in the nineteenth century, have much in common with the more recent personal and humorous narrations of Linklater and O'Hanlon.

In pointing to these differences between different kinds of travel accounts, we are aware of an important distinction, although not an absolute one, between travel for purposes of scientific exploration or for official commercial and administrative reasons, and travel for individual motives: curiosity, adventure, excitement, or endurance. It is perhaps appropriate at this juncture to relate the variations in travel writing to the historical context within which it occurred. Again, although there is not a perfect correlation, we can determine certain tendencies. This will also enable me to sketch out briefly some basic information on Borneo and the main phases of European contact and interest.

Europeans first came to the East Indies in search of commercial opportunities in such activities as the Oriental spice trade. Initially, the Portuguese and Spanish established contacts with the Muslim sultanates in South-East Asia from the early sixteenth century. Subsequently, the Dutch, British, and French followed; the Dutch were especially dominant in commerce in the seventeenth century, the British and the French increasingly from the eighteenth century.

Borneo was not a focus of special European interest, but the Europeans did establish commercial relations there. However, early contacts were quite superficial. Borneo was commonly described in the context of a long voyage of discovery and survey. The focus of attention was the opening up or expansion of trade with the coastal Muslim states. To a Portuguese, Spaniard, or Italian of the sixteenth century, or even a Dutchman or Englishman of the seventeenth century, these Islamic port centres must have appeared impressive and sophisticated places. Pigafetta's description of Brunei in the early sixteenth century conveys this sense of Oriental luxury. The early voyagers by sea merely anchored for a while at certain ports of call, and we can see that Beeckman, in his account of early eighteenth-century Banjarmasin, was providing various sorts of information about the island of Borneo, particularly the interior regions, based on reports and rumour.

Nevertheless, as we begin to approach the mid-nineteenth century, with the flowering of industrial capitalism in Europe and the assertion of Western supremacy over South-East Asian peoples and territories, the tone and detail of travel accounts change. Observations in the context of a sea journey are still important: witness Marryat's account of the voyage of the HMS *Samarang*. But now we begin to get more direct contact with the interior native populations. Even Marryat and his companions venture to Dayak longhouses and witness pagan festivals and customs. As the European presence becomes more firmly established, head-hunters and coastal raiders increasingly pacified, and an administrative structure, initially supported by military means, put in place, European travellers can be more ambitious. We still witness Charles Brooke, in the 1850s, embarking on military expeditions against warring Dayak tribes, but his uncle, Sir James Brooke, the First White Rajah of Sarawak, progressively provides the opportunities in his territories for European scientists and explorers to collect and discover in relative safety.

The great English naturalist, Alfred Russel Wallace, spends over a year in Sarawak in the mid-1850s, collecting, observing, and classifying. The Italian naturalist, Odoardo Beccari, travels more extensively in Sarawak and Dutch West Borneo in the mid-1860s. We are here presented with authoritative scientific accounts of the flora and fauna of Borneo, although Wallace was also interested in recording the particularities of local cultures. In addition, Spenser St. John achieves one of the first successful ascents of Mount Kinabalu in north-eastern Borneo in 1858, an expedition of scientific discovery, but also a demonstration of Victorian endeavour and courage.

It is not surprising, then, that as European dominance becomes increasingly apparent, and is exercised in political, military, and economic terms, the time arrives not only for a certain lightheartedness, but also for the rather more extreme expressions of European cultural imperialism in writings on travel. We are now confronted with the early tourists: those like Frederick Boyle, a Victorian gentleman, who visits Sarawak out of curiosity, and in search of amusement and adventure. Once amusement and curiosity are sufficiently satisfied, he departs. We find in Boyle's narrative all the evidence for European perceptions of their superiority and of their privileged position in evolutionary terms. Natives are to be observed, classified, and given trinkets. Even some of the more serious travellers, like the Norwegian, Carl Bock, who explored south-eastern Borneo on behalf of the Dutch colonial government in the late 1870s, were preoccupied with the exotic traits of savage peoples. Bock sadly had been duped by folk-tales and rumours in Borneo of cannibalism and men with tails. But given the scientific environment of evolutionary theories, it is hardly surprising that certain late nineteenth-century explorers and scientists were excited by the possibility of discovering the 'missing link', creatures which combined human and animal characteristics. Borneo, of course, was an ideal hunting-ground for these travellers: Daniel Beeckman, as far back as the early eighteenth century, had already established the orang-utan, 'the man of the forests', as a creature of fable.

Nevertheless, as we come to the end of the nineteenth century and into the first forty years of the twentieth century, when European colonialism is at its most secure, the attitudes of travellers and explorers change yet again. Colonial policies, directed to the improvement of the living conditions and welfare of the natives, give rise to a mixture of paternalistic views about the dependent populations, and attempts to appreciate alien cultures in their own terms. Charles Hose, as a long-term resident and scholar-administrator in Sarawak, was interested in the sympathetic study of local cultures in their own right, but also as a means to better administer them; Carl Lumholtz, the Norwegian explorer, travelled in the south-eastern and central regions of Dutch Borneo, seeing in natives the qualities which he believed Westerners had lost; Hendrik Tillema recorded indigenous traditions in Dutch East Borneo before they disappeared, and argued for changes in Dutch policies towards native health and hygiene. Yet, all these writers still embraced notions of the importance of European protection of native populations and held the view that colonial governments had responsibilities towards them.

This was patently not the case for some Europeans in the period immediately prior to independence. One is struck by the egalitarian attitudes which characterize the writings of Tom Harrisson and Malcolm MacDonald. Harrisson very much desires to understand the indigenous populations on their own terms without hint of condescension; and MacDonald, although a colonial officer in the dying days of

British post-war imperialism in South-East Asia, adopts a relaxed, easygoing relationship with local people.

These attitudes carry through into the period of independence. The later travellers write about Borneo people as people. Of course, as travellers in foreign lands they still dwell on the strange and exotic, but now the narrative is much more personal. It is as much about the thoughts, feelings, and experiences of the traveller as it is about the surrounding peoples and landscapes. The dispassionate scientific accounts of the nineteenth and early twentieth centuries give way to the personal adventure. When there is not much more to discover, one discovers oneself. What one gets in the recent accounts of Hanbury-Tenison, O'Hanlon, Linklater, and Hansen is a much more intimate, and usually therefore amusing portrait of the Western traveller in physically and culturally alien environments.

The fortitude and sang-froid of the Victorian travellers and some of the late colonial explorers and adventurers become the weakness and ineptitude of the recent traveller. St. John, Lumholtz, Tillema, and others shrugged off the difficulties and demands of travel in tropical rain forest environments. The O'Hanlons and Linklaters to some extent succumb to them. Lumholtz is in command of the rain forest; Hansen, some sixty-five years later, appears, at times, a lost child in it. One is struck by the self-deprecating tone of the recent literature. The certainty of nineteenth- and early twentieth-century Europeans, backed by the power of the colonial state, becomes the uncertainty of the travellers of the later twentieth century in the period of national independence. We would not expect St. John, Lumholtz, or Tillema to ask, 'What on earth am I doing here?' We are really not surprised when at times O'Hanlon, Linklater, and Hansen are doubtful about the wisdom of their journey.

The extracts in this anthology demonstrate many of the different dimensions, motivations, and experiences of travel. We are able to see how Borneo has provided for many travellers both the raw material for scientific exploration and observation as well as the opportunity for personal discovery and reflection.

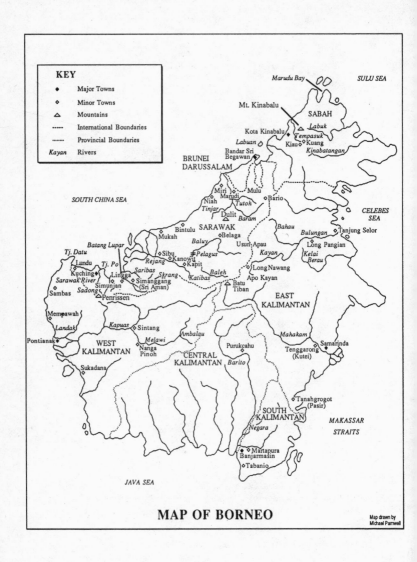

MAP OF BORNEO

KEY

◆   Major Towns

◇   Minor Towns

△   Mountains

-----   International Boundaries

·······   Provincial Boundaries

Kayan   Rivers

SULU SEA

Marudu Bay

Mt. Kinabalu

SABAH

Labuk

Kota Kinabalu ◆   *Tempasuk*

Kiau ◇ Kuang

Labuan ◇

Kinabatangan

Bandar Sri
Begawan ◆

BRUNEI
DARUSSALAM

*CELEBES
SEA*

SOUTH CHINA SEA

Miri ◇   Mulu

Niah   Marudi

Tinjar   *Tutoh*   Bario

△ *Dulit*

Baram

Bintulu ◇   SARAWAK

Mukah   Belaga *Bahau*   *Bulungan*   Tanjung Selor

Baluy

≠ *Pelagus*   Usun Apau   Long Pangian

Tj. Datu

Lundu   *Tj. Po*   Sibu ◇   Kanowit   *Kayan*   *Kelai*

Rejang   Kapit ◇   *Berau*

Kuching ◆   *Saribas*   *Baleh*

Sarawak River   Lingga ◇   *Skrang*   *Katibas*   Long Nawang

Simunjan ◇   Simanggang   Apo Kayan

Sadong   (Sri Aman)   Batu

Sambas   △ Penrissen   Tiban

EAST
KALIMANTAN

Mempawah ◇

Landak   *Kapuas*   Sintang ◇

Pontianak ◆   *Ambalau*

WEST   *Melawi*   *Mahakam*   Samarinda ◆

KALIMANTAN   Nanga   Purukcahu ◇   Tenggarong ◇

Pinoh   CENTRAL   (Kutei)

KALIMANTAN   *Barito*

Sukadana ◇

◇ Tanahgrogot
(Pasir)

SOUTH
KALIMANTAN

*MAKASSAR
STRAITS*

Negara

◇ Martapura
Banjarmasin ◆

◇ Tabanio

JAVA SEA

The Wonders of Nature

1
The Beauty of the Rain Forest

PATRICK M. SYNGE

Patrick Synge, of Corpus Christi College, Cambridge, was a member of the Oxford University Expedition which spent some six months in Sarawak from June 1932. This scientific expedition, under the auspices of the Oxford Exploration Club, was funded by the British Museum and various leading scientific organizations, such as the Royal Geographical Society. There were eight members of the team, 'all youthful' Oxford and Cambridge men. Six of the participants provided chapters for Borneo Jungle: *Tom Harrisson, expedition leader and ornithologist; John Ford, general zoologist; A. W. Moore entomologist; Edward Shackleton, surveyor; C. H. Hartley, ornithologist; and Patrick Synge, botanist. The remaining members were B. M. Hobby, entomologist, P. W. Richards, botanist. The book was written and compiled five years after the expedition.*

The team concentrated their studies in the area around the foot of Mount Dulit in the Tinjar basin, a tributary of the Baram River. A major task was collecting specimens of fauna and flora. A notable achievement was Edward Shackleton's first ascent of Mount Mulu, Sarawak's highest mountain. Unfortunately, Patrick Synge developed septicaemia in October 1932, caused by poisoned leech bites. He had to go to the coastal town of Miri for treatment, and he did not return to the Tinjar. Instead, he spent a brief time orchid-hunting a short distance from Miri at Niah.

1

*Harrisson says, in his brief introduction to Synge's chapter:
'Living always in deep greens and teeming tropical life, the
Bornean natives see the whole world, both of reality and
dream, in terms of twisting, tendrilous, exuberant vitality.
Creeper and carving, tree stem and tattoo mark, orchid and
art-forms—these are inseparably related in the pattern of
jungle life. To Synge, too, they are inseparable in his interest
and understanding. So, here, first he takes the jungle, then the
art' (p. 151).*

*Synge's original contribution was divided into two parts:
the 'beauty of the jungle' and 'the beauty in the tribe'. In the
following extract, Synge captures the 'frenzied luxuriance'
of the rain forest.*

FORESTS, so luxuriant that they seemed a frenzy of
greens, a solid wall, a never-ending skyscraper of leaves
overhanging the water's edge; behind, darkness and
mystery and more and more plants, many still unknown and
all thrilling to the young botanist.

A new land, my first expedition to the tropics and this
superb forest country—giant trees like the grains of sand on
the seashore, their trunks hidden by the mass of foliage, the
only visible stem generally the rope of a liana hanging in
graceful festoons. Then, four thousand feet above, a moss
forest, where striped and gaudy pitchers hung on twisted
stems and white and yellow orchids scented the air, all
backed by this feather-bed of moss; everywhere, on the
ground, on the tree trunks, on the branches—caves of green
moss, covering tussocks and boughs, often twisted and con-
torted into weird shapes and bearing the semblance of
strange faces.

I never felt alone in the forests. Surely it is possible for
such a frenzied luxuriance to convey a thrill to the traveller,
an enjoyment in such careless plenty, a feeling of unity with
life around, everywhere vigorous and healthy.

As we moved slowly up river, these forests were like a
dream come true to me, a dream of abundance, of beauty
and peace, but also of mystery alive behind this wall. All
abstract words for concrete objects—yet the power of the
forest forces is so great that we cannot be impartial to it. Man
cannot fail to be dominated by the forests.

Borneo is dominated by its vegetation; all the instincts and life of the people reflect it. To understand the life of the Borneo peoples, we must ourselves experience the great forest, the shapes of the leaves and the curves of the twisting lianas. The people practise shifting cultivation and only scratch a small part of the surface of the land, burning and clearing very roughly an area in which they plant hill rice along the banks of the rivers: after a few years the padi (rice) field is deserted, and the ground rapidly becomes covered with secondary forest even denser than the primary forest which it has replaced. Probably hundreds of years must pass before the big trees grow again and primary forest is restored; no one knows for certain how long. This primary forest is the real old forest, distinguished by the abundance of very large trees and the relative thinness of the undergrowth; in secondary forest there are few large trees, and the undergrowth and tangle of lianas is much denser. It is generally possible to walk through primary forest without cutting a path; this can seldom be done in secondary growth.

One of our objects in Borneo was the investigation of the structure and species of the primary forest, of which there is little now left on the Tinjar banks. Away from the rivers, however, there is an abundance which seems unending. Looking down on the Tinjar Valley from our High Camp on the ridge of Mount Dulit it was possible to distinguish a sharp line of division between the primary and secondary forest—a line which ran roughly parallel to the river bank and from half a mile to a mile away from it. The secondary forest is bright green and looks not unlike a smooth lawn; indeed, this resemblance is one of the chief dangers to aviation in the tropics; I believe there are records of pilots mistaking such forest for smooth ground and attempting a landing. The primary forest is a dull dense green in colour, and the crowns of individual trees can be seen projecting above the general level even from some distance.

So dominant is the forest that it is said to be possible for an orang-outang to travel from the south to the north of Borneo without descending from the tree-tops. His only barrier would be the big rivers and, since the majority of these run north and south, they would merely prevent his spread longitudinally east and west.

Botany and plant-collecting in the tropical forests are not quite the same as botany in England. In Borneo we started with the belief (subsequently found not to be quite true) that everything floral was unknown. Still, the plants were less known than the animals. We wanted a complete collection—everything from trees to mosses. We discovered over thirty-three new species of orchids alone and very many new species of other groups. Instead of setting out with a vasculum and a knife we started with porters, large baskets and axes. We formed quite a procession every morning, headed by Richards and myself. Then came Ngadiman, an excellent trained Malay collector, whose services had been lent us by the Director of the Singapore Botanic Gardens. He was a little man, and he always wore a most immaculate white topee. Ngadiman was followed by Lumbor, a forestry guard, lent by the Rajah;* then came two or three Kenyahs or Punans carrying baskets and axes on their backs, never on their heads like the Africans. If the forest was thick, and we strayed from the main path, they would cut a path with their parangs; always when we left the main path they would cut and bend over small twigs so that we should be able to find our way back again. It is very easy indeed to become lost only a few yards from the path.

In camp we left Omar, a valuable old Malay, who dealt most skilfully with the Herbarium material. He would change the papers and keep a fire going nearly all day drying the specimens; in fact, he almost cooked some of them with excellent results.

There is something about the rain forests of Borneo that will always lure back the traveller who has once visited them; the luxuriance, the chaos of tangled growth and the vastness of some of the trees inspire awe. A tropical forest is not dark, as is often suggested; the sunlight does penetrate through the 'canopy', and the ground-level is neither so dark nor so gloomy as in a thick pine wood in Europe.

Many imagine that these deep tropical jungles form a paradise of large and brilliant blossoms. The flowers indeed are there, but they are not found in conspicuous masses as on an English heath or in an Alpine meadow. If you want to see hillsides covered with colour, go to the Alps or the Himalayas, and not to the tropical rain forest, which is indeed a paradise, but a paradise of greens; there is so much of this

*Rajah Vyner Brooke.

colour and so many different shades that the effect is some-what unreal. Often I felt that our wanderings in the forest must be only a pleasant dream and that I should wake to find myself in Cambridge, slumbering over the fire.

In the forest we had to search hard, frequently with field-glasses, to discover the flowers; it is easy to go for a long dis-tance without seeing any at all, for they are mostly on the tops of the trees where the light reaches them; only a few fallen corollas or petals below betray their presence.

Our path was not a smooth or flat one, just a trail cut through the forest on the mountain-side. Practically all the ground between the river and the mountain consists of ridges with narrow and sometimes knife-edged tops, up and down which we perpetually scrambled, slipping and sliding on the clay soil and clinging to the frailest supports to help us up; indeed, sometimes I felt quite glad when we found a tree in flower and were able to rest while it was examined. The main crest of Dulit consists of an escarpment of Lower Miocene sandstone which runs more or less parallel to the general course of the river for some twenty miles, maintaining a ridge of fairly even height. In many places, where it was very steep, the Kenyahs had prepared for us ladders made out of notched tree trunks; sometimes these ladders were almost vertical for twenty or thirty feet, and it would have been impossible for laden porters to ascend the mountain without them. The dampness natural to the mountain and the forest soon covered these trunks with a mass of minute green plants and algæ and made them extremely slippery. Since there was often no hand-rail we had to balance ourselves precariously on them. We quickly found that ordinary sand-shoes with rubber soles were much the best kind of shoes to wear, and that nailed boots slipped the worst on the ladders.

There was a small stream which wandered down from the escarpment of Dulit into the Tinjar by our camp. Our trail crossed this stream eight times. Generally it was easily ford-able by stepping-stones, but after heavy rain it would become a raging torrent and form an effective barrier between the two camps. One day, for botanical purposes, we felled a large tree across one of the fords, which we thought would provide an adequate bridge. The trunk was several feet clear of the water. A few days later it was gone, and we found that it had been swept by the spate several hundred yards downstream into a quite useless position.

Along the stream we could see the size of the forest trees since it was possible to view them from a distance. Many were as much as a hundred and fifty feet in height and a hundred feet to the first branch. Near the base the trunks were often buttressed, and the buttresses would sweep out in serpentine curves from the base of the tree. The cause of the buttressing is still largely unknown to botanists, although it is probable that they do provide some extra support for the tree.

Along the stream, also, we could see best the luxuriant growth of liana and epiphyte. I think that these are two of the factors which contribute most to the beauty of the forest. The stems of the lianas would hang like bell-ropes from the tops of the trees to enter the ground far below. Lianas are a characteristic of tropical forests. They are woody climbers with rope-like stems and masses of luxuriant foliage tumbling and smothering in the tree-top level where the light is reached. They are most useful to the wandering botanist, since up them a man can generally be persuaded to climb, exactly as up a rope, to gather specimens from the tree-tops. In England the nearest approach to these lianas is the Traveller's Joy, *Clematis vitalba*, but it displays little of the rope-like stem and the great masses of luxuriant growth found in tropical lianas.

Many of these ropes assumed weird forms. We saw one which resembled closely a corkscrew eighty to ninety feet long, winding in a spiral up to the forest top. It was probably a species of *Bauhinia*, a leguminous liana common in the tropics. Another such climber had a flattened stem twisting in graceful curves and loops, but always repeating the same series of forms and sequence of curves, even as some ornate wallpaper returns to the same form again and again.

The rattans or climbing palms caused us much trouble. The commonest was *Calamus*, a climbing palm with beautiful feathery fronds, but in which the leaflets at the ends of the fronds and the stems bear stout hooks; these would frequently catch in our clothes, our hair or our flesh. The stem of this palm often attains a length of several hundred feet, and when stripped is largely used by the people as rattan for fastening round joints in their houses and making baskets, etc. Although slender, it is very strong. It is the rope of Borneo as banana fibre is the rope of East Africa.

The high temperature and the high humidity of the tropical forests encourage growth of all kinds; but to obtain light it is necessary for a plant to grow up quickly to the top of the forest as a young tree does, or else to grow on some support such as the branch or the trunk of the tree. These are the epiphytes—orchids, aroids, ferns, mosses and many other groups. They are not parasitic on the trees, but are merely supported by them. Round their roots they collect leaf-mould, and from the damp air they draw in moisture. This is especially true of many of the epiphytic orchids whose roots are surrounded by a layer of spongy tissue called the velamen, which acts like a sponge.

Epiphytic on the leaves of many plants, including other epiphytes, there would often be small masses of liverworts; merely a green form to the naked eye, but when looked at under a lens a delicate and beautiful structure of slender branches and green plant body. These epiphyllous liverworts seemed to me to represent the acme of epiphytism, the last word in sponging on your neighbour.

We had been asked especially to look out for and to collect as many of the flowers of the trees as possible. Often the flowers of the trees are the last flowers to be collected in the tropical forest, since they are not often visible from below and are difficult to obtain. Whenever a few flowers were found on the ground, we would settle down to a 'tree conference'. This became an established part of our routine and would last from five to twenty minutes. It provided for me an unfailing source of amusement. It was no easy matter to decide from which tree the flowers came.

Everyone would peer upwards into the 'canopy'; field-glasses would be passed from hand to hand until the canopy had been inspected from every angle except the one vital one—namely, from above. Ngadiman and Lumbor would go round, blazing all the trees, examining the latex, smelling the bark and sap. At last one of them would come to a decision; nearly always they were correct. We ourselves could seldom see any flowers above.

If it was possible for a man to climb the tree or by a neighbouring liana to reach the flowers, this was done. We had one man who was a champion climber. I am sure that he would have put many of Britain's best gymnasts to shame. But if the flowers could be obtained in no other way, we would cut down the tree. Axes would be pieced together; the head was generally carried separate to the shaft; the men would start work and down the tree would come, often in an amazingly short time, as the axes were very small and the wood of many of the trees very hard. The wood of bilian, one of Borneo's finest timbers, is so heavy and close-grained that it will not float in water. Always the fall of a big tree fascinates me. In these Bornean forests it was magnificent, since the tree-tops were so thick that one tree could not fall without disturbing others; for several moments after the main crash, little pieces would continue to fall, while the sound echoed round and round from tree trunk to tree trunk; following on by contrast came a deeper silence than that to which we were accustomed, until gradually the smaller noises of the forest began to assert themselves again. One forest giant would generally bring down several smaller trees and lianas in its wake, and we left disappointed if we could not collect considerably more than the flowers of the actual tree for specimens.

Systematic science demands many sacrifices in the way of fine trees, flowers, birds and animals, beautiful butterflies and strange insects, but in the case of the tree, the fall of one giant provides light and space for seedlings to grow up and his place is filled again.

While Richards measured the trunk of the tree for diameter and height, and took out a longitudinal section for a timber specimen, I would scramble about among the débris and collect epiphytes, particularly orchids, of which we obtained some interesting species in this way. The epiphytic flora of a great forest tree is enormous. I have often thought that it would repay an ecological study. Along one branch alone there may be hundreds of plants, and with them are collected humus and mosses and insects dependent on the plants. A single tree-top is like a small world of its own.

There are, however, a few trees in the tropical forest which do not produce their flowers at the top branches, but out of the old wood of their trunks often only a few feet from the ground. Such trees are peculiar to the tropical forests. I don't think that any occur in England or in temperate countries.

The almost invariable association of the larger epiphytes with ants caused me much discomfort, and it seemed that the smaller the ant the more painful was the bite. Luckily the effect only lasted for a short time, but frequently I had to suspend my collecting while we brushed and picked the ants off. There are several epiphytes which produce specialised internal cavities in which the ants live, such as *Myrmecodia*, a plant in which the base is formed into a large tuber, honeycombed with passages inhabited by ants. There is also a *Polypodium* fern which has a great flattened base hollowed into chambers. It has been suggested that the ants protect the plant from other insects, but it seems more likely that their main importance to the plant lies in the humus which they collect. There are also species of *Macaranga*, plants belonging to the spurge family, which contain hollow cavities for the ants in the flower-stalks. There is a small hole by which they can enter and leave the cavity. The exact relation between the plant and the ant is not known.

We were particularly keen to find orchids which might prove of some horticultural value, and indeed several that we found were of considerable interest for their bizarre form and delicate colouring. I found them more interesting by far and sometimes more beautiful than the gigantic and showy hybrids produced in such numbers to-day. For pure show, however, few of the Bornean species could compare with the products of modern orchid hybridisers. The orchid flora is very rich, and undoubtedly many species yet remain to be recorded. Richards, who had also visited British Guiana, reported that the flora of that region seemed even richer in orchids of horticultural value than the flora of Borneo; this would easily be confirmed from an inspection of any orchid exhibit at a flower show, since all the *Odontoglossums* and the *Cattleyas* are American in origin.

The prevalence of epiphytes, particularly orchids, seems to be closely related to the roughness of the bark. In the case of a very smooth bark little humus or moisture collects on the branch and few epiphytes are found, while in the case of trees with rougher bark the higher branches are often covered entirely with epiphytes which take advantage of the humus and moisture collected by the rough bark.

Scent is a frequent characteristic of the Bornean orchids. There was one small *Coelogyne* with white and yellow flowers common in the moss forest. It had a very sweet smell. I could always tell when a plant was near by this means, and would then search until I found it. There is another Bornean orchid, *Bulbophyllum Beccari*, which is reputed to have a smell far stronger and far more unpleasant than that of any durian, which can be smelt fifty yards off. The flowers are reddish-brown and are, I imagine, fly-pollinated. Nearly all plants which are pollinated by flies smell like bad meat. We have only to think of *Aristolochia* or *Stapelia* to confirm this. The leaves, also, are magnificent. It is probable that they are the largest of any orchid leaves, being entire and over eighteen inches in length by twelve inches in width. The stem is almost wooden and ascends the tree trunk, clinging with small roots like ivy, while the vast leaves grow out as brackets, collecting a mass of leaf-mould at their base.

In extreme contrast to this plant are the minute flattened stems of *Taeniophyllum*, which we frequently found on the highest branches of the trees. Here there are no leaves, and the stems resemble slender green tapes creeping over the branches. The flowers also are minute.

On a tree trunk a little distance from the trail, I found one of the finest orchids we saw in Borneo. The flowers were large, about two inches across, and borne singly one on a stem. They were pale buff brown in colour, with orange flecks and lines. The lip of the orchid was poised and curved forward, so that it moved in the wind; the lateral petals were like two superb orange-brown moustaches, emerging above a high, stiff collar. There was a large mass of this plant. We took half, and I still remember Ngadiman walking down the mountain with it, supported like a wreath round his neck. Subsequently this orchid flowered in England in Sir Jeremiah Colman's famous collection and received an Award of Merit at the Chelsea Flower Show in 1938 under the name of *Bulbophyllum Lobbii var Gatton Park*. It was unfortunate that the name suggested rather a hybrid raised at Gatton Park than a species collected on Mount Dulit in Borneo. I do think that such varietal names should not be given to species collected in the wild, but that all names should as far as possible be either descriptive of the plant or be derived from the place whence it came.

As well as these fine orchids, we found a number of species of *Cirrhopetalum*, with deep crimson, reddish-brown and orange flowers, small hooded flowers, from the base of which extended a long strap-shaped labellum. These flowers were grouped together at the end of a long stem, so that they often formed a semicircle of colour a couple of inches across, delicately poised and swaying in the wind.

Grammatophyllum, the largest orchid in the world, grows in Borneo. We found a clump on the edge of a tree by the rice field clearing. At first I thought it was a kind of palm, so vast were its long filamentous leaves, borne on a thick stem often six or eight feet in length. Unfortunately we did not see it in flower here, but I subsequently saw it in flower in the Botanic Gardens at Buitenzorg in Java, perched in a tree of the great *Canarium* avenue. The flowers are orange and brown, several inches across, and borne on great six-foot spikes, while the leaves must have been an equal length. Occasionally plants are seen in England, but I have heard that it very seldom flowers. Another of the astounding orchids of Borneo is *Arachnanthe (Vanda) Lowii*, whose flowers are borne on long garlands often thirty feet in length. The flowers are crimson and brown and about two inches in diameter, but the two basal flowers of each spike are absolutely different from the others, both in colour and shape. They are bright yellow and have small crimson spots. This wonderful plant is named after Sir Hugh Low, an early resident magistrate who made large collections.

All these orchids of the forest require to be grown in a stove-house in England, but the orchids of the moss forest into which we suddenly entered at four thousand feet may well be grown cooler.

Although many of the orchids are epiphytic on branches or trunks of other trees, they are not parasitic. We found many interesting parasites, however. Once our collector found buds of the giant parasite Rafflesia, the largest flower in the world, but unfortunately he cut them together with the root of the liana on which they were growing. Severed from their host, they did not open, and we never found any more.

Perhaps the most brilliant floral sight in the forest was presented by the flowers of a member of the *Loranthaceae*, a parasite allied to the English mistletoe. The flowers had a brilliant red and orange tubular perianth and covered the whole crown of a great tree with colour.

Some of the most interesting trees of the mixed forest were

the 'strangling' figs, the seed of which is carried by birds and germinates in a fork among the top branches of some tall tree where the fig grows epiphytically. Long roots are produced which often clasp the trunk of the 'host' tree and kill it in much the same way as ivy kills trees in England. They reach the ground and enter the soil as normal roots; thus by the time the 'host' tree is dead, the fig is independent. A bizarre and fantastic appearance is frequently presented by these roots, which are twisted and gnarled, while a forest of aerial roots (which grow straight for one hundred feet from tree-top to ground) is produced when the tree gets older.

Perhaps, more than any other plant, palms are associated with the tropics, and Borneo was full of beautiful palms. As we came up river in the brackish estuaries, the banks were lined with the Nipah palms, wonderful great fronds, thirty feet of glorious feather, springing directly out of the mud, waving in the breeze, glowing orange in the evening light. Then in the forest there were the rattans, aggressive climbing palms; occasionally we saw one of the beautiful sago palms, dignified trees with a twenty-foot trunk surmounted by great waving plumes. The sago of the pudding comes from the pith. In the actual undergrowth of the forest we would find the Licuala palms, plants with no central trunk and large fan-shaped leaves. These were used often for roofing. Our High Camp was roofed with them. It did leak in heavy rain.

Ferns were also abundant in the forest, ranging from the great clumps of *Angiopteris*, akin only to the primitive monster, to the filmy fern, so ethereal that its fronds were transparent, being often, I believe, only one cell thick. The birds' nest fern generally perched precariously, attached to the side of tree trunks, the fronds surrounding a perfect nest.

Among the fungi we found many closely resembling English species. One of my most exciting finds, however, was a species of *Dictyophora*, fantastic and exotic in its appearance. The base resembled a delicate hair-net, pale pink in colour, suspended from a central fleshy stalk which ended in a porous yellow 'head' apparently secreting some nectar, since even large insects were continually attracted to it. Ten inches in height, this fungus was a magnificent sight, almost a surrealist object, growing out of an old prostrate tree trunk.

Above 3,600 feet we suddenly entered a different world. It was the moss forest, a weird, a fantastic zone, where everything was covered deep in green cushions and tussocks of moss sponges of green, moist, squashy, frightening but beautiful. The transition from the 'mixed' jungle type of forest to the 'moss' forest is quite abrupt and presents one of the problems which the results of the expedition have not so far explained. At the bottom of the ladder up a small cliff was ordinary forest, at the top was moss forest. Drifting cloud is probably a factor in the formation of the moss forest, but no certain correlation can be made of the sudden change between the two types of forest and the cloud-level. It also seemed to us to become suddenly cold as we entered the moss forest. Dr Hose indeed compared the weather in the Dulit moss forest to a bleak November morning in England. The effect was accentuated by the contrast with the heat below, and we were all glad to wear sweaters and often tweed coats as well.

The 'moss' forest consists of small trees, very few of which are more than forty feet high: they present a fantastic appearance, being gnarled and bent and covered in the lower parts with a dense growth of mosses and liverworts, hanging in thick mats and long festoons from the trunks and the undergrowth; although the forest is called 'moss', liverworts actually predominate. The growth is generally a foot in depth, and in some places is as much as ten feet, giving the appearance of green grottos or fairy caves, forming arches and wreaths from branch to branch, while green pillars and stalactites join floor to roof. Pitcher plants, orchids and rhododendrons grow here in profusion. Many are new to science, and they give more colour to the moss forest than there is in the forests of lower levels, where the chief source of colour was the mauve and purples of the young leaves, drooping delicately and bashfully from their stems.

There is a curious contrast between the luxuriance of the moss with its general dampness and the leaves of the trees, the majority of which are small and often ericoid (heather-like) or thick and leathery in form. It would appear difficult to correlate such apparently sclerophyllous features—features of plants growing in very dry or very acid environment—with the prevailing humidity, although it seems that the reasons must rest in factors of soil and light and acidity. But this last

13

factor is more a result of the moss than the cause of it. In general, more light penetrates into the 'moss' forest than into the 'mixed' jungle forest. There are great contrasts in this respect which add to the mystery. The sclerophyllous and ericoid type of leaf is particularly conspicuous on the tops of the peaks where the exposure to sunlight is greater and the growth is dwarfed, while the plants are slightly different from those of the general moss forest.

In the early morning the dewdrops sparkle on the feathery mosses, and the filmy ferns are almost ethereal in their delicacy; then it does not require much imagination to conjure up little winged figures peeping out of the grottos made by the moss. There were mosses growing here on the ground which might have been miniature Christmas trees, hung with fairy bangles when the dew flashed with light on them. The pitchers would have provided drinking-cups, nestling in the moss or swaying in the wind.

It was indeed an enchanting place, although there always seemed something mysterious and uncanny lurking among the fantastic shapes of the trees and moss, and it does not seem strange that many of the Borneo peoples regard the mountain-tops as the home of the spirits of the dead.

In the 'moss' forest flora, Australian types, such as *Casuarina Dacrydium* and *Leptospermum* predominate, while in the lower forest, Malayan types are easily dominant. It is pleasant to speculate that the 'island' mountain summits of Borneo are relicts from the day when there was a land-bridge between Asia and Australia—islands which survived the inundation of the lower lands which now surround them. I fear, however, that such a statement can be little more than a speculation. There is Wallace's line of deep water just east of the island, while there is no such line separating Borneo from Malaya. This moss forest differs completely from the lower rain forest both in structure and in the actual plants; indeed, out of the several thousand numbers of plants which we collected, so far only fourteen species have been found common to both the lower rain forest and the moss forest.

Nestling against the tussocks of moss were the pitchers of Nepenthes—Nepenthe, the old goddess of sleep and oblivion—and certainly it is oblivion for the many insects which find their way into the pitchers and are drowned there and slowly digested—for the Nepenthes are insectivorous

14

plants. But they are more—they are beautiful plants; their pitchers are streaked and painted with a theatrical brilliance, their form designed by a Cellini endowed with a Machiavellian and wholly diabolical cunning.

As we emerged into the moss forest we came upon a pool into which a little waterfall continuously poured, throwing out endless ripples to the green edges. Near here we found our first moss forest pitcher—*Nepenthes tentaculata*. It was one of the smallest, but one of the most beautiful. The pitchers are borne at the ends of the leaves dangling on a curved stem, which is a prolongation of the mid-rib of the leaf. It is hard to describe their form; they are like some very elaborate and exotic pipe, coloured on the outside green and streaked with crimson. They are variable in colour. Often they have a bluish-purple tinge. The inside of the pitcher is pale blue. Often six inches in length and two inches in diameter, they hold a considerable amount of liquid. The angles and the lid are feathered with deep crimson hues. Always these plants grew in shade and seemed to like a lower light intensity than the other species we found, which clambered up to the light.

In Borneo there is no doubt about their insectivorous habits, but it is doubtful whether the insect food is so necessary to them. In English greenhouses they seem to grow quite well without any insect food. Perhaps we may regard the decayed insect food as a savoury titbit, supplying extra nitrates and other salts, to use an anthropomorphic simile. In nearly all species the rim of the pitcher is smooth and extremely slippery so that any insect, attracted by the bright colour or by the honey secreted by the glands around the rim, would be inclined to fall down into the fluid below. Some of the larger pitchers contain more than a pint of fluid. One-way traffic only is ensured as the rim of the pitcher is generally formed after the manner of one of those unspillable ink-pots and has stiff hairs projecting downwards from the edge. There is also a lid which does not close but against which any insect which attempted to fly out would be likely to collide.

The inside of the pitcher is covered with small glands which secrete a fluid, allied to the proteolytic enzymes of our own insides and possessing digestive properties. The fluid inside the pitcher is distinctly acid, and it is probable that the digestive functions of the enzyme can only work in an acid medium. Even a young, unopened pitcher contains some acid fluid. In a large pitcher the greater part of the fluid is probably due to water which has condensed inside the pitcher or entered as rain. Inside the pitcher are remains of many kinds of small insects in varying stages of decomposition, mostly small flies, beetles and moths. Occasionally, parts of large insects are found inside the pitchers. But, an amazing fact, the pitchers also contain a considerable fauna of living aquatic insect larvæ, particularly mosquito larvæ. It seems probable that the digestive enzyme can only act on dead matter, and even then its digestive powers are weak, as it is considerably diluted with rain-water. It has also been suggested, and there is some experimental evidence to support the suggestion, that these mosquito and fly larvæ contain an anti-protease substance which would inhibit their digestion by the proteolytic enzyme of the pitchers.

As we cut our way through the moss forest we found other remarkable species of pitcher plants, *Nepenthes Rheinwardtiana*, *Nepenthes Veitchii* and *Nepenthes stenophylla*. I think that *N. Reinwardtiana* was the most graceful and beautiful of all that we met. The slender stems scramble up through the moss and small trees to reach the light. Often they are thirty feet long, and at the top only are found the large crimson pitchers dangling free in the air, ten or twelve to a plant. As the leaves die off, so do the pitchers, and more grow above on the young leaves. In shape they resemble narrow flasks and are often as much as fourteen inches in length. They are not streaked and blotchy as other species, but a uniform rich deep crimson in colour. Inside the pitcher is a pale green in colour, while just below the cup are two brilliant emerald spots which gleam as eyes.

The pitchers of *Nepenthes Veitchii* are large, and resemble both in shape and colour the popular hybrid often seen in cultivation, and named after Sir W. Thistleton Dyer. It is a magnificent plant, a flamboyant beauty. The pitchers are covered thickly with a down of pale pink hairs, while the lip of the mouth is prolonged upwards into a fan-like structure of extreme slipperiness coloured with brilliant diagonal stripes of green and scarlet. They are borne on rigid stems which adpress them closely to the tree trunk. The stiff leaves also clasp closely round the trunk.

16

Nepenthes stenophylla was distinctive of the slightly more exposed and drier situations on the very top of the various summits of the Dulit range. The pitchers are large, graceful in shape and brightly coloured, streaked with raw-meatlike crimson and scarlet. At the apex of the pitcher where it joins, the mid-rib is always a little spiral twist, somehow comic and delightful.

In the moss forest we also found many beautiful orchids. The most common was a *Coelogyne* which proved to be a new species. It had white and yellow scented flowers in a long pendant raceme which hung down on the moss like a necklace. It was so common in some parts that we could often smell the flowers before we actually rounded the corner and found the plants. We also found a fine *Dendrobium* with large white and yellow flowers and a most curious terrestrial orchid, *Corybas Johannis Winkleri*, with a small deep crimson and white cup-shaped flower from which protruded a crimson lip shaped like a ledge, while from the rim of the cup extended three long filaments like whiskers. This grew out of the base of a single heart-shaped and beautifully-veined leaf.

My first day in the moss forest I got a great surprise. Looking up at the fork of a big tree I saw a brilliant orange flower. I sent Arwang to fetch it, and he came down with one of the most brilliant epiphytic rhododendrons that I have ever seen. Later, we found another kind with fine shell-pink flowers and yet another with scarlet flowers. They all grew as epiphytes and were rhododendrons of the type we sometimes see in hot-houses labelled Javanese hybrids. But they were as fine as any of their type I have seen in cultivation. We collected some seeds, but unfortunately they did not germinate in England. On the mountains of New Guinea there are also fine rhododendrons perhaps even finer than these, and from there it is conceivable that some might be hardy in English gardens. Some might even be tolerant of lime, since many of the New Guinea mountains are made of limestone with innumerable knife-edge ridges.

From the moss forest we passed over the ridge of Dulit down into the forest by the stream of the Koyan. This was quite different in appearance to that we had found in the Tinjar, and we found many new plants.

The most unusual feature of this forest is the soil, which is almost pure sand, and undoubtedly this is the limiting factor in controlling and forming this type of forest. A curious feature about this type of forest was the absence of buttressed trees, although with such a light soil there would appear to be a greater need for them than in the mixed type of forest where the soil is a heavy clay or loam. The ground-level here gets more light, and there is a much more dense undergrowth. There was a definite tendency towards species dominance, and *Agathis alba* the dammar produced thirty-five per cent of the trees over sixteen inches in diameter. This tree, a broad-leaved conifer, sometimes reaches a considerable size; it produces a valuable resin which is collected by the people and exported for use as a colourless varnish.

Richards told me that he had found a somewhat similar type on sandy soils in British Guiana. The species and genera of the plants were different, but the general appearance was similar. In the same way there was a certain similarity in appearance, although not in species, between the moss forest which I saw here in Borneo at four thousand feet on Mount Dulit and that which I saw later at ten thousand feet on Ruwenzori—the Mountains of the Moon—in Central Africa. There also was the abrupt change of vegetation as we passed from one zone to another on the mountain.

In this Koyan sand forest we found more pitcher plants, and they were in many cases the same species as we saw in the sand forest by Marudi and in the scrub sandy country inland from Miri.

Nepenthes bicalcarata was perhaps the largest and the most magnificent of all the pitchers we met; in size it is only rivalled by *N. Rajah* and *N. Lowii*, the largest pitchers known. They are found on Mount Kinabalu in North Borneo. The pitchers are nearly globular and often were as much as six inches in diameter across the mouth. They vary in colour from a pale green to a deep crimson. It is a vigorous species, and several plants were found fifteen feet in height. The most exciting part of the pitcher is two stout spines which project downward from the lid and are very prominent. There is a story that one eminent botanist, desirous of hoodwinking the public, affixed a dead rat to these spines and proclaimed the plant as a mammal-catcher. Normally, however, insects form its only prey, while it is fairly certain that the plant can grow quite successfully, without any insect food, living as other plants do. Here also were the red-streaked *N. rafflesiana* and the paler green *N. leptochila*. In the swamp part of the Marudi forest we also saw *N. ampullaria*, a species which has clusters of small pitchers in groups one above another round a stout stem ascending to the forest top. These pitchers had no lid, but a small straplike handle sticking out from the side. They would have made splendid cups. These plants were growing in a swamp forest where the trees formed acrid roots sticking up from the ground and covered with warty encrustations.

Seeds of some of the species of pitcher plants collected in Borneo have been germinated successfully in England, but their growth is very slow. The atmosphere must be uniformly damp, but different species seem to like different amounts of light. From the extreme localisation of their distribution in the field it would seem that Nepenthes are very closely related to their environment and that any successful grower must follow these conditions as far as possible. I doubt also whether the species from the sand forest habitat and from the moss forest should even be grown together in the same house. At the Singapore Botanic Gardens I was told that they could not grow mountain species of Nepenthes successfully, but had to send them to the garden on Penang Hill.

In describing tropical rain forest, the terms 'stratification' and 'canopy' are frequently used, not only by botanists, but also by novelists and the more sensational travel writers; in their application to the 'mixed' forests of Borneo, it seems that they are somewhat misplaced and indeed incorrect. Stratification suggests a definite break between the layers, a discontinuity, while in the 'mixed' forests there was no definite break at any level nor was there any sudden change in the gradient of temperature or humidity from the tree-tops to the ground. The undergrowth and the trees tend to arrange themselves in somewhat vague layers, but there is a gradual and not a sudden change from one layer to the other, since much of the lower layers is composed of young plants of the species which, when mature, occupy the higher layers. The only layer which could possibly be called a 'stratum' is represented by the crowns of the tallest trees about a hundred and ten feet in height, the base of which was often separated from the tops of the crowns in the lower layer by a gap of a few feet. These crowns were not in contact laterally and so did not present a continuous 'stratum' or a 'canopy' which suggests a closed and a flat covering. Nevertheless it was easy for an orang-outang to jump from one to the other. Numerous flecks of sunlight penetrated through to the forest floor, while the undergrowth was very uneven, doubtless due to this reason. The crowns of the second layer of trees, which average about sixty feet in height, are frequently in contact with their neighbours and intertangled with lianas, and this layer has the best claim to be called a 'canopy', but it is neither completely closed nor is it confined to any definite level, but grades imperceptibly into the layer of small trees below, while the undergrowth layer is continuous with this. This idea of structure was gained by clear felling and measuring the trees on sample strips of forest. The ornithologists also report that there was no sort of distinct stratification of bird life in the forest.

The botanical procession normally started out about eight o'clock and walked and worked till lunch-time. We would eat lunch seated on some old log or on the ground. Lumbor would often manage to cook a hot meal, lighting a fire with amazing dexterity, although all around the wood appeared to be damp.

One of the men would prepare for us a small cup constructed out of palm leaf and made into the shape of a box. As a drinking-vessel it proved excellent. We would return to camp generally about four, and then would come the serious work of laying out and writing up notes on our specimens. On to each a number had to be tied corresponding to the number in the note-book, in which was entered details about habitat, growth, abundance, colour of flowers, etc.

In the stream by our camp there was a most charming bathing-pool where we would often swim of an afternoon or evening, apparently quite secure from crocodiles. I also used the stream, but a little lower down, for developing photographs, finding the water cooler and cleaner than the water of the main river. Photography was an important part of our job. There I spent many evenings before dusk sitting on a rock by the side of the torrent and meditating, while at intervals shaking the developing tank.

Returning alone to the camp I would generally be rewarded with the sight of more animal life than when I moved with a party. Great hornbills could sometimes be seen flapping about in the tree-tops, while a small squirrel would peer out of the undergrowth at me, or a vast squirrel, with a brush like a fox, would hurry across the path. I would also be cheered by a series of fairy-lights, dancing at different levels in the forest. These were the entomologists' light traps, which were suspended from several large trees in the neighbourhood of the camp. They hung in chains, each trap at a different height, and it was hoped in this way to gain some idea of the zonation of insects from the ground to the tree-tops.

Many gifts of plants and insects were brought to us by the Borneo peoples, and the majority of these, particularly the insects, were striking and bizarre; it will be a long time before I forget Hobby's noble exclamation of 'OH, JOY', when a moribund and damaged specimen of the commonest insect was brought to him in the middle of lunch and required immediate attention. The man was invariably gratified by such acclamation of his gift and would then bring another next day, generally at the same time. I fear the botanists could not rise to such hilarious gratitude; yet we tried to show our appreciation, frequently with the gift of small quantities of coarse tobacco. A few good plants were obtained in this way, but they were generally accompanied by insufficient data as to habit of growth and location.

Once, expecting a flower, I was presented with a small water-snake in a paper bag. The situation was, however, my fault since I had failed to understand the name given to the contents of the bag, which had been introduced as 'ular bungah'. 'Bungah' I knew to mean 'flower' and so expressed a desire of the gift, unfortunately failing to comprehend the 'ular' or snake part. However, when the bag was opened, I managed to move hurriedly to another part of the boat, and the snake jumped overboard, where it swam about quite happily amid general laughter. I was not really sorry to see it go, since I had no material for preserving snakes on the trip.

Through all my forest wanderings flitted the most beautiful butterflies that I have ever seen, and *Ornithoptera Brookeana* was the finest of them all. Moving through the tree-tops and over the edges of the cliff they seemed more like small birds than butterflies, as indeed the Latin name implies. Their wings are covered with peacock blue-green scales, which shimmer and gleam in the sun. This insect seemed like a flower that had taken to the air, a symbol of the many

21

beautiful things of Borneo, the native bird-winged butterfly of the white Rajah Brooke.

'Beauty in Borneo', in *Borneo Jungle. An Account of the Oxford University Expedition of 1932*, Tom Harrisson (ed.), Lindsay Drummond Ltd., London, 1938, pp. 153–77.

2
All on the Bornean Shore

CHARLES HOSE

Charles Hose, born in Hertfordshire, England, in 1863, was one of the most famous and distinguished administrative officers of Rajah Sir Charles Brooke, the Second White Rajah of Sarawak. Hose began his studies at Cambridge in 1882, but abandoned university life in 1884 and took up an administrative cadetship in Sarawak. He served the Raj for twenty-three years, mainly in the Baram region among the Kayan and Kenyah Dayaks. He was made Officer-in-Charge of the Baram in 1888 when he was still only in his mid-twenties. Three years later he became Resident of the Fourth Division, the administrative region mainly comprising the Baram basin. In 1904 he was transferred to Sibu in Sea Dayak or Iban country as Resident of the Third Division.

While in Sarawak he developed a deep scholarly interest in its natural history and ethnology, and he was a formidable collector of Bornean artefacts and specimens for the University Museum at Cambridge and other institutions. Although he had never been formally trained in these matters, Hose acquired an international reputation for his studies of Bornean nature and culture.

He retired from service in 1907 on a pension, and returned to England. From then until his death in 1929 he gave lectures and wrote several major books. His most well-known scholarly text, co-authored with William McDougall, is the two-volume compendium, The Pagan Tribes of Borneo *(Macmillan, London, 1912). He was also awarded the Honorary Degree of D.Sc. by his old university, Cambridge, in recognition of his scholarly endeavours.*

The following extract is Chapter 1, 'All on the Bornean Shore', from his Field-Book of a Jungle-Wallah. *It dwells on the coastal fauna and flora of Borneo, and provides an appropriate complement to Synge's description of the interior rain forests.*

I awoke with the dawn and, for a moment, wondered where I was; for the early morning in the tropics has a romantic element, bringing recognition and arousing mystery. Here was I, in my own familiar corner of the world—anchored in a small sailing vessel at the mouth of a minor river on the Bornean coast; I knew very well what was going to happen, yet I was prepared for small surprises. So true is it that Nature, which is ever the same, makes all things new.

Looking through my mosquito-curtains and waiting for the coffee which my servant, with Chinese precision, was preparing, I watched the opening of day. On the shore the mist was lifting from the woods and the dew stood out clear and bold on the undergrowth. Around me I heard the cry of awakened beasts, the whistles and calls of many birds, the high treble of insects, many varying voices; beneath them all the ground-bass of the infinite Sea.

On a sandy point, a hundred yards or so away, my Chinaman is conjuring back life into last night's fire; elsewhere, the boatmen are preparing their morning rice. To supplement it, one of them has collected some *Umbut*, the unopened heart-leaves of the Nibong palm; another, more enterprising, can be seen returning along the beach after a successful search for turtle's eggs. These are a delicacy easily prepared; one boils them like any other egg, pulls off the leathery shell, and consumes them at one's ease, with kindly thoughts of the creature that produces them, for she is perhaps the most interesting reptile of the Bornean coast. There are two kinds of turtle found here, one the Green Turtle (*Chelone mydas*); the other, the beautiful, but smaller, Hawksbill which yields the tortoise-shell and is known as *Chelone imbricata*. Green turtles are found mostly on the small islands and sandy shores of the coast; their movements are so calculated and pawky as to suggest a Scottish ancestry. Very often a prospective mother-turtle takes a long and devious road to her nesting-place, looking carefully about her before she lands. Well beyond the tide-line, she scoops out a hole in the dry sand, using her flipper like a hoe; and, having laid her eggs, returns to sea by an entirely different route. The egg-hunter, therefore, finds himself playing a game of 'Hunt the Thimble', with no one to say 'Hot' or 'Cold'.

After the hatching period, which is about five or six weeks, the babies slowly work themselves out of their sandy birth-place, and scramble down to the sea in long processions; before they reach the deeper waters where they live, many are devoured by sharks.

The turtle is a tactician, but Man, taught by experience, sets his wits against him; and in Celebes, the next-door island to Borneo, it is interesting to watch the local methods of catching them as they lie asleep on the surface of the sea, or in shallow water when grazing. A native with a rope tied round his waist dives from a boat into the water and clasps the creature in his arms; when he has got a good hold, his companions haul at the rope and bring in the two together. An even more spectacular method is one in which a small fish, *Echeneis naucrates*, having a flat sucker on the top of its head, is employed. The fish is tied tightly by the tail with a piece of fine string and dropped into a likely spot for turtle, usually in shallow water. In its efforts to escape it swims about until it lights on a turtle, to which it attaches itself by the sucker, probably in order to get a purchase. The turtle having been thus located is then collected in the ordinary manner.

Breakfast finished, we start on our journey along the coast. Malays are excellent sailors, as indeed are all peoples of Arab descent; and it is quite thrilling to watch our small ship, skil-fully handled, take each incoming wave so that nothing more than a light spray comes on board. When we reach the open sea the going becomes easier and we hoist a couple of mat-ting sails, about eight feet square, on our bamboo mast. Before we finally leave the river, however, and while the water is still fresh, we notice, among the decaying vegetation near the banks, a number of curious spouting fishes (*Toxotes jaculator*). These are a kind of perch, and get their name from their habit of stunning their prey (flies and other insects) by spouting water at them from a short distance, as they rest on an overhanging leaf. Hereabouts, especially at the mouth of a sandy river-beach, one may catch a sight of a King Crab

(*Limulus*), or *Blankas* as the Malays call him. He is a strange antediluvian-looking creature, of a warm grey colour shot with green, roughly half-spherical in shape, with an extraordinarily tough shell. His tail works on a hinge and looks like a marlin-spike or a cobbler's awl; it is used for purposes of entrenchment, for this curious beast loves a comfortable dug-out in the sand, into which it settles for protection from the heat. This tail is also used to ward off enemies, and makes a formidable weapon, although there is no poison emitted into the wound. When he is properly settled in his lair, you can see nothing of him but two of his eyes. Of these organs, the King Crab is blessed with, seemingly, two pairs, anterior or median, one on each side of the raised part of the foremost spike of the carapace, which are only visible on careful inspection, and two more posterior and lateral. It seems that only one pair of these lateral eyes is used, the median having ceased to function.

The creature is said to be a native delicacy, but personally I have only tasted the mass of small eggs which, varying considerably in size, are found (of all places in this world) in what may be described as its forehead.

Dr Gordon of the Natural History Museum acquaints me with the curious fact that the ovaries are found in the fore part of the shell or carapace of the King Crab, more correctly called the Sea Scorpion, about which very little appears to have been recorded.

In both sexes the gonads (reproductive organs) consist of systems of ramifying tubes lying in the shield above the other organs of the body. The males are smaller than the females. At the breeding season these creatures come ashore in pairs during high spring tide, the male clinging to the shield of the female. The eggs are fertilised after they have been deposited in the series of holes which the female digs for their reception. The first hole selected is near the upper limit of high tide, and the mass of eggs produced is said to be half a pint in volume.

At the mouth of the river are floating a number of great logs carried down from the back of nowhere, and riding idly on the swell. Some of these are of soft wood, and, being churned by the ocean during the north-east monsoon, turn to pulp, and form, together with various kinds of sticks and decaying vegetation, a compact mass, not unlike

25

the leaf-mould of English gardens. For this the Bornean native has a peculiar use, employing it in a sort of Turkish bath, known as *Bertangas*, for the treatment of rheumatism. The patient is enveloped in a close-fitting mantle of this material, which covers his whole body leaving his head alone emerging from a repulsive-looking cone. Heat is now introduced from below through bamboo tubes to start a sweating process; the patient meanwhile looking very much like Mr Cyril Maude in 'Lord Richard in the Pantry'. As a specific this treatment is considered invaluable, and I am inclined to believe that there must be some sort of iodine salt in the decayed vegetable matter. It has a not unpleasant odour, with a tang of the sea about it reminiscent of our seaweed.

The logs of the harder woods remain more or less unravaged, except for the depredations of *Teredo navalis*, or the ship-worm, with whose burrows many are completely honey-combed. When one looks at one of these creatures (which one can do quite easily, as the logs are alive with them) one wonders how it can create such havoc. The *Teredo*, who is known in Borneo as *Temilok*, did not always live this fixed sort of life; at one time in its existence it must have been a free-swimming creature, having special organs of its own with which to guard itself against the dangers of the deep. The larva swims in all directions with extreme agility; later the external skin bursts, and, after being encrusted with calcareous salts, becomes a shell, which is at first oval, then triangular, and at last very nearly spherical. The larva of the *Teredo* possesses, moreover, sense-organs similar to those of several molluscs, and also eyes. In Eastern waters an enormous species called *Teredo gigantica* is to be found which sometimes attains the length of five to six feet and a diameter of two to three inches. This species, however, does not bore in timber or make its habitation in logs, contenting itself with boring into the hardened mud of the sea-bed.

One is rather inclined to wonder what use Providence can have designed for the *Teredo*, until one finds that most coast natives regard them as a delicacy. In the East the faculty of wonder soon dies from overwork—especially in matters of diet; lizards, sea-slugs, shark's fins, locusts and the grubs of flying ants are eaten in many places—but in Borneo and the neighbouring islands one comes across something more *outré* still. Personally, I do not see myself eating these nasty-looking worms, but they figure on the menus of the local gourmets, fried like whitebait, the head and hard shell-like jaws alone being removed. I have often wondered, in view of the damage done by the *Teredo*, whether this custom is not part of a policy of Hate; 'You eat our wood,' says the Bornean, 'Very well! we will eat *you.*'

About mid-day, set high on a cape, we notice a lighthouse, and I think of what one of Kipling's characters, in this very part of the world, says: 'You understand, us English are always looking up marks and lighting sea-ways all the world over, never asking with your leave or by your leave, seeing that the sea concerns us more than anyone else.' But there are other people concerned with lights besides ourselves; and the fascination which a bright light has for fish, is turned to advantage by the coast-people, when searching for cray-fish in these shallow transparent waters. They stalk the creatures with a sort of barbed spear about seven feet long, which they push in front of them along the bottom. In order to make manipulation easier and keep the spear more or less stable, a sort of strut is attached at an angle and reaching down to the sea-bottom; while about two feet up the shaft and just above water-level, is fixed a candle which serves a double purpose, attracting the crayfish and guiding the fishermen.

Fishermen other than human seem to have discovered this use of a light, for there are certain creatures known as 'Angler-Fish' which use this method. Of these fish, for a description of which I am indebted to Dr Tate Regan, FRS, of the British Museum, there are about fifty different kinds, exhibiting many varieties of form and size, but uniform in this, that they all have the first ray of the dorsal fin placed on the top of the head and modified into a line and bait. One group of these, known as *Ceratioids*, live in mid-ocean, about half-way between the surface and the bottom. In these gloomy waters their bait is a luminous bulb, either set directly on the head, or else at the end of a long filament, in some cases projecting from the skull and looking like a rod and line. In one species (*Lasiognathus*) the line actually extends beyond the bait, and ends in a triangle of hooks. Here indeed is the Complete Angler.

Perhaps, however, the most remarkable peculiarity of these

Oceanic Anglers is that all the free swimming-fish are females, and that all the males are dwarfed and parasitic on the
females. The habits and conditions of life of these fishes, solitary, sluggish, floating about in the darkness of the middle
depths of the ocean, make it evident that it might be difficult
for a mature fish to find a mate.

This difficulty appears to have been overcome by the
males, as soon as they are hatched, when they are relatively
numerous, seeking the females, and when they find one,
holding on to her for life.

The males first hold on by the mouths, then the lips and
tongue fuse with the skin of the female, and the two fishes
subsequently become completely united; the male being
nourished by the blood of the female and the blood system
of the two being continuous.

The Ceratioids are unique among back-boned animals in
having dwarfed males, and in having the males nourished in
this manner.

As we pass along, clumps of Casuarinas and Screw-pines
stand out boldly; while down the bare cliffs tumble glistening
amber-coloured waterfalls. I always love stopping at one of
these falls, for the water gives a thrill like nothing else in the
world and both soothes and invigorates; after it one can the
better enjoy the rich colours of the seascape, and (to be
materialistic) the pleasures of lunch. As we watch our make-
shift sail bringing us nearer to the chosen spot, our steersman
(I always make a point of having the same man, when possible) points to a shoal of porpoises, or perhaps a shark,
following a few yards in our wake; the shark is probably as
ready for his food as we are. Overhead the white Sea-Eagles
and Brahmany Kites whirl in great circles, every now and
then darting down at the sight of a large fish, each vying with
each in the grace of their sweeping motions. On the shore
the locals (optimists, surely) shout and wave their hands at
them, trying to make them drop their prey.

Lunch over, one walks leisurely along the shore, where, if
shrimping is in progress, the curious sight may be seen of
millions of tiny shrimps passing along in line through the
shallow water. As the rays of the hot sun shine down through
the water, the whole seems to have changed to a beautiful
pink. When I first saw these numerous tiny pink prawns, I
was reminded of the artist who painted the live lobster red

and wondered if I myself was falling into the same delusion. However, on examining some, I found that the bodies of the prawns were of the ordinary greyish-brown colour. Their legs, however, were red and there are so many of them that, perhaps together with refraction, they tinge the whole mass of water a delicate pink.

At certain seasons, the natives catch them in close-meshed nets fixed on a pole, much like our shrimping nets, which they push to and fro in the shallow water along the beach. Basketfuls of them are spread out on mats and dried in the sun. They are eaten with rice, or else pounded up with salt into shrimp-paste; for a month or so in every year the whole atmosphere reeks with the smell of shrimp-paste while this favourite Malay article of diet is being prepared.

A similar Malayan delicacy, but one which appeals to Europeans, is a larger kind of prawn (*Palæmon*), found in the fresh or brackish water at the mouths of most of the rivers. This creature is almost as large as a lobster and has two very long blue claws. It is caught either in box-like traps made of bark, or else in cast-nets—in either case a ground-bait is used consisting of grains of rice mixed with some tasty and strong-smelling preparation. A piece of stinking decayed coconut flesh (copra) placed in one of these lobster traps proves a most attractive bait.

Along the rocky foreshore one often sees beautiful yellow ground orchids of the Calanthe family, growing on the sides of the cliffs and apparently without much soil. They have a beautiful spike of golden flower rather like a spiræa, and large flat leaves.

As one wanders along these wild, rocky coasts with their beautiful caves and luxuriant plant life it seems almost impossible to associate their majestic beauty with scenes of a time when pirates ravaged the seas, and killed and warred against the peoples who inhabited the shore, or captured and sold them into slavery. Yet it was in the creeks and rivers of this northern coast that the notorious Illanun pirates of the eighteenth century 'dug themselves in', a pest so formidable that the British Admiralty considered it 'certain death' to venture into their strongholds; and it was here and hereabouts that the famous Rajah Brooke and the 'Little Admiral', Sir Henry Keppel, earned the thanks of civilisation, smoking them out as one might a nest of hornets.

Pondering on this past history I sat down to think. But my thoughts were disturbed, for I noticed that two beautiful birds (Bee-eaters, or *Merops*), which had been flying to and fro about me, had alighted on the ground near by. They are migrants and are known to the people of Borneo as *Burong taun*, which means 'the Annual bird'. Their colouring is, in the main, green; but the neck, the upper back, and the wings have a broad colouring of ruddy brown; the lower part of the back is buff, and the long middle feathers of the tail are tipped with black. The forehead is pale green and white, and the throat bright yellow.

I watched the pretty things for some time, and then turning my eyes for a moment to a vessel out at sea, found they had disappeared. I had been looking in the same direction all the time and should have noticed had they flown away, so I got up and walked slowly to the spot where I had first seen them, and, to my astonishment, discovered several small holes in the ground, from one of which, while I watched, a bird flew out. When one considers the beak and the feet of these birds, one is not inclined to think of them as capable of making holes; however, picking up a piece of stick I pushed it down the hole to see how far it would go, suspecting that these birds were nesting. The hole ran parallel to the surface at a depth of a few inches for a distance of a foot or so, and at the end I discovered a nest with five glossy white eggs. These eggs were almost exactly like those of a Kingfisher. A little later I noticed that the Casuarina trees all about seemed to be covered with these birds, apparently mating.

In the forest regions of the Hinterland there is another species of Bee-eater, a fairly large bird, about the size of a blackbird. Its plumage is green, with a pinkish-red head. I have never been able to find where the forest variety nests, but as I watched these little coast birds of the same genus I could not but think it likely that, since the sea-coast Bee-eater nests in a hole in the ground, like the Kingfisher, his brother of the forest must build his nest in the hollow of a tree or a rotten log, or, possibly, in the ground, where the roots of a fallen tree may have exposed the soil.

Continuing our journey slowly along the coast, we observe a number of small caves of sandstone formation, similar to the limestone caves found farther inland only rather smaller. Here a wealth of vegetation seems to spring from nowhere but solid rock. Along the shore, adhering to the walls of the caves, may be seen the nests of the swift known as *Collocalia linchii*. This is a small bird, mostly black in colour, with a white-speckled chest, which builds a nest of moss and straw, smeared over with some gelatinous substance by which the nest is made to adhere to the rock. Owing to the nest not being entirely of this substance it has no important commercial value to the natives, who collect other edible swifts' nests for the Chinese market. This little swift is considered rather a nuisance by the natives; for he is a tyrant and has a habit of frequenting the haunts of the two other species of swifts, the *Collocalia lowii* and *Collocalia fuciphaga*, and disturbing them, occupying the nests and more suitable positions and ousting the more valuable species.

Along marshy plains which sometimes occur on the coast, and which are undoubtedly long-forgotten clearings of the forest, short sedge grasses may be seen growing. Here one finds the Drosera, a carnivorous plant not unlike the two European species found in the fenlands of East Anglia. The little plant is about the size of a two shilling piece and has the habit of catching flies and insects in its leaves. Another fairly common plant with similar carnivorous habits is a species of Pitcher-plant (*Nepenthes*), more or less a dwarf variety of the large beautifully-coloured jungle species. What was of greater interest was a species of Myrmecodia, a strange plant which seems to have struck the natives as something weird, as they call it *Anak hantu*, or 'child of the spirits'. At the basal portion of the stem it has a sort of hump, capable of storing dew and other moisture. The skin of this growth is so thick that it prevents evaporation, and, if the stems are cut across, they are found to be full of tiny catacombs and channels in which ants constantly shelter, breeding in prodigious numbers and making their permanent home there—for which reason the plants are called 'hospitating plants'—and it would seem that the biological connection which exists between them and the hospitating ants is a case of symbiosis for reciprocal advantage, the plant apparently entertaining the ants because they are useful as a defence against other insects and various hurtful creatures.

Quite a number of good-sized trees are found about here, growing right down to the beach; among them the *Hibiscus tiliaceus*, not like the 'shoe-flower' that one sees in the

Scillies and other sub-tropical climes, but a great tree as tall as a house, and with a trunk eighteen inches or more in diameter. It is an interesting tree, for from its inner bark the natives obtain a useful string known as *baro*, that procured from the younger branches being superior as the bark is more tender and pliant. Although in considerable use locally, it has never yet been turned to commercial use to any extent; but probably, in judicious hands, it might prove a valuable commodity. The large flower of this Hibiscus is of a beautiful yellow, which turns to a ruddy tinge as the tree grows older.

Before turning away, for we had to move on towards our evening's camp, I came upon a specimen of that extraordinary creature, the Ant-eater, or Pangolin (*Manis*), in the act of devouring a nest of red ants, on the bough of a tree. Having lost its teeth through long ages of disuse, this creature has a very long tongue with enormous salivary glands, the sticky secretions of which enable it to pick up the ants. Its chief means of defence are its sharp strong claws, which are used to tear an ant-hill to pieces. Its back is covered with a mass of hard horn-like scales. It is, however, able to emit at will an extremely unpleasant odour, enough to baffle the most hardy pursuers.

My friend, the late Robert Shelford, in his book, 'A Naturalist in Borneo', repeats an amusing Malay legend (which may or may not be true) about the Manis. When there is a shortage of his natural prey, he lies down in the forest, curls himself up into a ball, and shams dead. Soon thousands of ants, acting perhaps on a sort of 'bazaar rumour' of the jungle, flock together to feast on the supposed corpse, and swarm under the slightly raised edges of the Ant-eater's scales, in order to get to the soft skin underneath. 'When the Manis considers that it has collected sufficient numbers of the ants, the corpse comes to life again, straightens itself out, and in doing so shuts down the scales and imprisons the ants. It then trots off to the nearest pool of water or stream, into which it plunges and arches its back, thus raising the scales again. The ants float off on to the surface of the water and are licked up with the long slender tongue.'

I once captured one of these creatures and put him in a large wooden box, which I nailed down until I could find him a more permanent residence. My night's rest was broken by the noise made by him trying to escape; and next morning

I found that he had used his claws to such effect that he had torn away about half the lid, and had escaped.

The Ant-eater is a slow mover on the surface, but a quick burrower; it makes its home in holes in the ground or in hollow trees. In Borneo we have the long-tailed species only (*Manis javanensis*). He is probably a remnant from a prehistoric age, and, I am bound to say, he looks the part.

On one of these coastal voyages, before making a start in the morning, I turned aside with a couple of my Iban followers, to examine a tall *Tapang* tree, the larger branches of which were literally covered with clusters of wild bees clinging to the underside, at least a hundred families. When I reached the foot of the tree, to my astonishment a half-grown honey-bear—about the size of a large bull-dog—came sliding down the trunk, tail foremost, growling fiercely, and using his powerful claws as a brake. The last few feet he accomplished by rolling; but he rose rapidly, growing angrier every moment, and then reared up on his hind legs as if he wanted to fight me. He reminded me rather of the way Mr Snodgrass 'announced in a loud voice that he was going to begin'; but he evidently thought the better of it, for he sheered off to behind the spreading root of the tree, and then with a gesture of contempt and a final snarl, ambled off into the forest.

When I looked round, I found that my Ibans had disappeared. I knew that they alone of the Bornean tribes hold bears in respect, but I asked them why they had fled. They explained that it was strictly *mali* (taboo) to injure a bear, or even to attempt to do so, as for instance by laughing at one. I asked them if they would eat bear's flesh, but they seemed horrified at the very idea. They look on the Bear as one of their omen-animals, and the Bear to most Ibans is as near an approach to Totemism as anything that I have ever found in Borneo; and indeed in their attitude towards the animal one finds Totemism in the germ. The Bear is, for some individual Ibans, a Gnarong, a sort of guardian angel, specially belonging to one person much like the Roman *Genius*, or the Egyptian God 'who walks by the side', or even the ($\delta\alpha\acute{\iota}\mu\omega\nu$) of Socrates; a personal helper and protector taking the form of a chance-seen animal. Now some Ibans have a tribal protecting spirit named Impernit, who, apparently, was an actual chief of the Undup river; for I have seen his tomb (*Rarong*, or *Sungkup*)—built above ground on piles—in which there is a tube inserted as a means of communication between this world, and the World after Death; here the chief's body lies, as the Ibans say, 'to be with us even in death'. Impernit was probably a great warrior some generations back; and the probability is that the Bear was the Gnarong of this tribal god, and is for that reason held sacred by the whole tribe.

The Field-Book of a Jungle-Wallah. Shore, River and Forest Life in Sarawak, H. F. and G. Witherby, London, 1929, pp. 1–24.

3
In Search of the Orang-utan

ODOARDO BECCARI

Odoardo Beccari, born in Florence in November 1843, was a great Italian naturalist and explorer. He studied botany and the natural sciences at the Universities of Pisa and Bologna between 1861 and 1864. He landed in Sarawak, still as a young man in his twenties, on 19 June 1865 in the company of the Marquis Giacomo Doria, a great patron of natural history. Beccari spent about two and a half years in Borneo studying the fauna and flora and collecting specimens. He eventually had to leave the island in January 1868. He had been struck by a fever, and on New Year's Day 1868 he wrote: 'My health ... had now completely broken down.... Fever attacks were now frequent and violent, and elephantiasis, which had shown itself some months before, was evidently increasing rapidly' (p. 351). (Many of these early explorers and scientists, travelling in the monsoon tropics, succumbed to fevers: in this collection, Wallace (Passage 9) and Lumholtz (Passage 12) are two further examples.)

Beccari travelled extensively in Sarawak, to interior Land Dayak areas, the region of Lundu, and the Iban country of the Batang Lupar. He also travelled to Labuan and Brunei, to the Rejang River and Belaga, and even went over the border into Dutch territory, journeying into the Kapuas Lakes area in West Borneo.

Even after his eventual departure from Sarawak, Beccari

continued travelling, and during the next ten years undertook expeditions to other parts of the East Indies, and to New Guinea and Ethiopia, returning briefly to Sarawak in December 1877 and then on to Sumatra. It was another two and a half decades before Beccari managed to put pen to paper and record his adventures in Wanderings in the Great Forests of Borneo. *It was first published in Florence as* Nelle Foreste di Borneo *in 1902, and then translated and published in English two years later.*

In the following extract, Beccari had journeyed to the Batang Lupar region in 1867. Unlike Wallace (Passage 9), another great naturalist, Beccari apparently had little interest in people. He was overwhelmingly preoccupied with natural history, and this extract gives us details of one of the most famous examples of Bornean wildlife, the orang-utan *or* maias (mayas).

The Earl of Cranbrook describes Beccari as 'a man of great intelligence, versatility and intuition, an indefatigable worker, austere and inflexible, courageous and stalwart, proud and almost misanthropic, not an easy character, not afraid of solitude, but, on the contrary, finding refuge in it. . . ' (1986: v).[1]

DURING the two years I had been wandering among the forests of Borneo, I had not yet met with a single Orang-utan; but up to that period botanical collections had so occupied my time, and the country I had explored had given me such rich results, that I had not cared to stray far from Kuching, where the great anthropoid ape is very rare, and to go in search of it on the Sadong or on the Batang Lupar, where it abounds.

On the Sadong Wallace had long resided and collected; I therefore chose the Batang Lupar, whence I could easily pass into the Dutch territory of Kapuas, and visit the lakes which exist along the upper portion of the course of that great river.

In March, 1867, the Tuan Muda,* having occasion to send his gunboat, the *Heartsease*, to Lingga, kindly allowed me to take this opportunity of going there with the larger portion of my provisions, while at the same time my men were to take

---

[1]'Introduction' to Odoardo Beccari, *Wanderings in the Great Forests of Borneo*, Oxford University Press reprint, Singapore, 1986, pp.v–ix.
*Charles Brooke (lit. 'Young Lord').

the sampan which was to convey me during the remainder of the journey.

At 8 a.m., on March 17th, the *Heartsease* left her moorings, steamed down the Sarawak river, and reached the sea by the Maratabas channel. The weather was splendid; the sea like a mirror. We turned eastwards, making straight for the mouth of the Batang Lupar. Behind us rose the dark bold outline of Tanjong Po, slowly emerging from the thin morning mist; and on our right the low coast line revealed itself with its monotonous fringe of verdure, consisting of mangroves where the shore is muddy, and of casuarinas where sand prevails. Behind this belt of interminable forest rises Gunong Lessong, remarkable for its truncated form and its wide base. Passing quite close to Pulo [Pulau] Burong, I could see that this little rocky island was completely clothed, especially on its upper portion, by a handsome palm, whose enormous racemes loaded with flowers and fruits looked like small Cypress trees protruding from the midst of crowns of sago-palm fronds. It is undoubtedly a species of *Eugeissonia*, which, although I was unable to examine it more closely, I consider identical with one I subsequently found on the banks of the Rejang and at Brunei. As it is a useful plant from which good feculum can be extracted, I should not be surprised if originally its seeds had been brought to the island by Dyaks. Borneo, forming the very centre of the area of their past piratical expeditions, may have been used by them as a victualling station. A little before sunset we passed the small island which stands in the middle of the mouth of the Batang Lupar. When the sun dipped below the horizon, darkness came on very suddenly; but the night was clear, and our captain being well acquainted with the soundings, we continued our way up stream. At 9 p.m. we had reached our goal, the old fort of Lingga, once the residence of the Tuan Muda, and now completely abandoned. It is placed on the right bank, near the mouth of the Lingga river, the first affluent to be met with on the right, ascending the main stream. As my boat had not yet arrived, I had my luggage taken into the fort—a low wooden building, hidden amidst coconuts and fruit trees. All around the soil was swampy and honeycombed by hosts of crustaceans, which make myriads of little hillocks with earth extracted from the burrows in which they live.

The next day, my boat still not having arrived, I took my gun and explored the neighbourhood. I was able to shoot several species of birds which I had not met with before; amongst them was *Lalage terat*, Cass., a bird which, in flight and size, is somewhat like a swallow. It has the habit of taking a few rapid turns in the air and then perching on the extremity of a bare branch of one of the trees growing on the banks of the river.

On the opposite side of the Lingga river the land is low, and was in former times occupied by rice fields, but at the period of my visit was overrun with a large kind of grass, a species of *Ischæmum*, which forms immense meadows, pleasant to see at a distance, but in which walking would be impossible, for it reaches a height of some eight or ten feet. Moreover, the soil underneath is a morass, and one would sink up to the knees in mud and slush. The mosquitoes thrive by the legion, and render life intolerable.

On the nineteenth of March I left Lingga fort before the tide flowed, but awaited the tidal wave at the mouth of the Sungei Batu, another affluent of the Batang Lupar, where, in a safe position, I was able to observe the curious effect that this produces in the shallower parts, where, instead of the ordinary bore, the water appears violently agitated in disordered movements, and seems as if it were boiling tumultuously. At 4 p.m. I reached Fort Simanggan without notable incidents.

The next day, about four o'clock in the afternoon, having ascertained that the tidal current had reached the fort, we continued our ascent of the river with its aid. Soon afterwards we passed the Undup, an affluent on the right bank, and later on the Sakarrang [Skrang], on the left. Higher up, the main stream, which still retains the name of Batang Lupar, grows much narrower. Up to this point the country could hardly be less attractive, with its low banks, bare and monotonous, or with at the most a few scattered trees. But these are the signs of a densely populous region, and of soil adapted to the cultivation of rice. The shrubs scattered over the country are the remains of forest species not entirely destroyed by fire during the clearings, and appear as strangers amidst the vegetation of the plains. We passed the night at Balassan, a Dyak village of nine families.

Early on the 21st we started paddling, aided by a slight tide for a short distance, but this was very soon overcome by the current of the river. I shot here a burong bubut (*Centrococcys eurycercus*, Cab. and Hein.), a large species of cuckoo, which keeps to open plains and abandoned rice fields, flying from bush to bush. Its loud and oft repeated cry—'bubu-bubu'—is heard for hours in monotonous regularity on these plains, and its native Malay name is derived from this peculiarity.

I saw here for the first time that singular fish (*Toxodes jaculator*) which has received from the natives the name of '*Ikan sumpit*', literally 'blowpipe-fish'. It is neither remarkable in shape nor coloration, but it has the strange power, on coming to the surface, of being able to squirt a jet of water from its mouth. This it uses with unerring aim against insects, such as grasshoppers, flies, and even spiders, resting on plants near the water's edge, causing them to fall into the water, where they become an easy prey to the clever marksman. The ikan sumpit has thus a special advantage over other fishes also preying on insects.

Primitive Man managed to obtain possession of living animals in motion by virtue of the admirable structure of his hand, which enabled him to grasp a stone or other missile, and to hurl it at the animal he wished to capture. Such must have been the origin of the first suggestion of implements of the chase. In Man's case the sentiment which caused the action was desire, followed by an act of volition. But it is indeed singular that a fish, intellectually so greatly man's inferior, should exhibit reasoning capacity similar to that of a human being under like conditions. The remote ancestor of the ikan sumpit must have beheld objects which it desired to possess, but which were beyond its means of capture, and, destitute of both prehensile organ or missile, may have tried to spit (if I may so express it) at the insect which, settled on a blade of grass overhanging the water, had tempted its avidity. The fish thus utilised the only means in its power towards an attempt to throw something at the desired prey. The conclusion is that acts of volition have induced the ancestor of the ikan sumpit to endeavour to perform certain movements in its buccal apparatus towards the attaining of an end for which originally its organism was not morphologically adapted. The modifications, therefore, which finally caused so perfect a water-ejecting apparatus to develop can only have had their origin in the stimulus I have indicated, namely, a voluntary act of the fish and the desire to get possession of an object which was useful to it.

The manner in which the ikan sumpit captures insects has much analogy to the methods of the chameleon. In both cases we have special adaptations in certain organs whose modification can only have been caused through impulses of the will. It must have been the wish to capture prey, and this only, that has rendered possible those morphological adaptations by means of which the desire could be attained.

It is, however, singular that, among the numerous series of its more stupid brethren, this little fish should alone have had, one far remote day, at the dawn of its specific existence, the spark of genius which led it to discover that spitting at a fly sitting beyond its reach would cause it to fall into the water and become an easy prey. It would thus appear that even in beings at present least gifted by intelligence, this latter can at one time have existed anterior to instinct, which in final analysis is merely an inherited form of intelligence.

We passed Bansi, a Dyak village-house containing nineteen families. The river banks continued bare and monotonous, but the mountains of Marop came into view. The only interesting plant I met with was a Loranthus (*Beccarina xiphostachya*, v. Tieghem), a magnificent species, parasitic upon a small tree hanging over the water, and covered with beautiful rose-coloured flowers five inches in length very similar to those of some of its congeners of the Andes, in which, however, the flowers are even more remarkable, attaining the extraordinary length of seven or more inches.

After a short rest at Unggan to cook our rice, we continued our ascent of the river, passing several Dyak villages. This is one of the more densely populated districts of Sarawak, and at the same time more cultivated, thus affording little to interest the botanist. The rocks I saw, and they were but few, were invariably sandstone. Towards three o'clock in the afternoon we reached the landing place for Marop. I disembarked my luggage at once, and stayed in the house of a Chinaman—there being quite a little Chinese village here. The following day, March 22nd, I found without difficulty Chinese and Dyak bearers to convey my luggage to Marop. The former did so by suspending the load, divided in two portions, at the ends of a bamboo pole resting on the shoulders; the latter carried their loads on their backs, secured with bands of bark passing under the package and over the head of the bearer.

The road from the landing place on the river to the village of Marop—about an hour distant—is one of the best I came across in Sarawak. One might even drive a light buggy or dog-cart over it, were such a conveyance known in these parts. I was delighted to see on the way that the primeval forest had not been all cleared away, and that there were places where it was evidently intact. It had, indeed, not a very vigorous aspect, but it looked different from that I was already acquainted with, which made me look forward to the possibility of finding some novelties. Meanwhile I came across a *Dipodium*, a ground orchid, with fair-sized, slightly perfumed flowers of a milk-white colour, covered with vinaceous blotches.

Marop is a Chinese village, placed in a small valley surrounded by low hills. The stream from which it takes its name runs through it, supplying an abundance of cool limpid water, and giving off a minor torrent which dashes merrily amidst the houses. The village was very clean; most of the houses were made with mats or palm leaves, but the big house, or residence of the Kunsi, the headman of the Chinese, in which I took up my quarters, was almost entirely built of wood. My lodgings were on a spacious platform forming a kind of first floor, where I made myself fairly comfortable, having ample room for my big and cumbrous cases.

I was impatient to explore the country; and as soon as I had seen my luggage safely housed, I made an excursion up the nearest hill, where I at once fell in with a troop of red-haired monkeys (*Semnopithecus rubicundus*), a fine species I had not met with before, as, like the orang-utan, it is not found in the neighbourhood of Kuching. In the afternoon I went up the Batu Lanko, the highest hill in the neighbourhood, though it hardly reaches the elevation of 300 feet. It owes its name to an enormous block of granite raised on other similar masses, so as to form a sort of cave or shelter ('batu' = stone, 'lanko' = hut). On the slopes of this granite hill I found layers of clay, evidently alluvial, with traces of gold. The spot was then abandoned, but from the disturbed condition of the surface over a large area it was plain that very active gold washing had gone on there not long before. The system followed is the usual one—that of washing the auriferous deposit in a stream of running water canalised so as to lead into successive flat pans or basins at decreasing levels, where the gold particles, on account of their greater specific gravity, remain, whilst the earthy and other lighter materials are washed away by the running water.

I extended my walk to Ruma Ajjit, a Dayak village, situated on the crest of a steep hill. Ajjit—for such was the name of the headman, or Orang-kaya, of the village—as soon as I approached him, took my right hand in his, and passed twice over my head a fowl which he held in his left hand. After this he presented me with the fowl, inviting me very civilly to sit near him by the hearth-stone. This was the place of honour, over which hung several smoked human heads, precious trophies of his past acts of bravery. He gave me siri and betel [quid], according to the established custom amongst Dyaks as well as Malays, the first act of hospitality towards a welcome guest; and after some conversation, having asked him to send me fowls which would be well paid for, and to get his people to collect animals for me, I took leave of my worthy Dyak chieftain and returned to my quarters in the Kunsi's house.

At Ruma Ajjit I saw an albino girl. She had a good figure, and in Europe might easily have been mistaken for a German or Swiss maid, with her fair hair, blue eyes, and full rosy face, but the latter was somewhat disfigured by scurfy spots and freckles.

On the twenty-third of March, with several Dyaks as guides, I again ascended Batu Lanko hill, where I had been told that orang-utans, or 'Mayas', as they are called here, had been seen. I did not meet with any, but found, and was able for the first time to study, their nests or shelters. The term nest is rightly applied to the beds or resting-places which these animals construct on trees wherever they remain for a time. They are formed of branches detached from the tree on which they are made, and heaped together, usually at a big bifurcation of the trunk. There is no attempt at anything like an arrangement, nor is there any roofing, and they merely form a platform which serves for the creature to lie down on.

The orang-utan nests I saw were evidently each for a single animal; possibly a united couple may build for themselves a more commodious couch, but I was unable to find out more of the domestic habits of these primates. As I have said, what I saw were merely beds or couches for lying down on; but I

'Orang Utan Attacked by Dyaks', from Alfred Russel Wallace, *The Malay Archipelago. The Land of the Orang Utan, and the Bird of Paradise*, Macmillan and Co., London, 1869.

think it very possible that on cold nights, or during rain, these creatures may also use branches and fronds as a shelter or to cover themselves with. It is well known that in captivity the orangs like to wrap themselves up in a cloth or blanket,

The forest in the vicinity of the village being deprived of most of its attractions, I directed my steps next day towards the low ground in search of plants, and was by no means unsuccessful. That evening all the sick and invalids of the village assembled at my house, for my fame as a doctor had spread far and wide. My system of cure was the simplest, and, thanks to my good fortune, gave splendid results. To those affected with fever I gave quinine; to those who suffered with dysentery, chlorodyne; to the others, fresh water, coloured with a little Worcestershire sauce. Sometimes I added a little arack; but I soon had to [be] careful with the latter remedy, for the number of my patients increased instead of diminishing.

My out-patients having all been attended to, I went to sit up for deer by moonlight in a 'lanko' which commanded a small plain surrounded by bushes, where the grass was very long and thick. The deer ought to have come here to feed, but the only thing that did come was clouds of mosquitoes, which, had I had any desire to sleep, would have effectually rendered it impossible, while, if they were not sufficient, the floor, formed as it was of large stakes, placed side by side, was not of such a nature as to tempt to drowsiness.

On the 25th I again went in search of plants towards the plain. From the hill I had noted all the localities where clumps of trees still stood, and each day I proposed visiting one.

Towards evening a Chinese hunter brought me the first orang-utan, but it was so mauled and covered with parang cuts that I did not skin it. Mayas were apparently far from being scarce in the neighbourhood of Marop, and I felt certain that I should soon be able to get better specimens. This one was a female of the kind named 'Mayas Kassa' by the Dyaks, who distinguish several varieties or kinds of the orang-utan. The hair on the body was red, the skin beneath was of a deep copper colour; the face was much darker—a blackish-olive.

Next day I went into the jungle in search of Mayas with the Chinaman who had brought me the one above mentioned.

Nevertheless, I was not favoured by fortune, and we wandered for four hours in the forest without seeing a single animal of the kind. When I got back I found another Chinaman waiting for me with a second Mayas, very similar to the first, but rather smaller. It was also a female of the Kassa variety, and it had still attached to it its little baby son, which had remained clinging to the mother when she fell wounded. In the fall the poor little creature had broken its left humerus. I prepared the skin of the mother, who had received a single bullet in the head, and had broken the bones of both arms in falling. None of my men were proficient in taxidermy, and I was thus obliged to do nearly all the work myself, to tell the truth, not too willingly. I had decided, however, to devote a whole month to orang-utans, and to preserve a complete series of these most interesting animals, both skins and skeletons, so I set to work at once without more ado. As I was eating my supper in the evening, the 'Tukan mas', or goldsmith of the village, came to tell me that he had killed a Mayas, but the hour being late had left it in the jungle. Three other Chinamen who were with him had remained on the spot, partly to guard it, and partly in the hope of shooting other specimens.

The Chinese at Marop were big and strong, and excellent walkers; they had come from Sambas, and were as well acclimatised as the Dyaks themselves. In the evening they used to gather round me and talk for an hour or two, asking me all sorts of questions on Europe and the Europeans, while some of their queries were, perhaps, somewhat less ingenuous than those of the Dyaks.

Next morning, March 27th, I finished preparing the skin of the Mayas which had been brought to me on the previous day. At noon they arrived with the one shot by the Tukan mas. It would have made an excellent specimen had it not been spoiled by the Chinaman who killed it, and who, in taking out the viscera, had badly split with his parang both the sternum and the pelvis. It was a male of the Mayas Kassa kind, and offered no appreciable differences from the female I had prepared already. I measured it carefully, with the following results:

Total height (vertex to soles of feet)	.	.	.	.	1.17 m
Across the outstretched arms		.	.	.	2.10 "
Trunk from vertex to coccyx .	.	.	.	.	0.73 "

Circumference of thorax below sternum (the viscera
    having been removed) .   .   .   .   0.81 ”

I may here state that I always took the measurement of the
height by stretching the animal on the ground and measuring
the distance between the crown of the head or vertex to the
under surface of the heel. The exaggerated dimensions of the
height of orangs, given, nevertheless, by conscientious and
trustworthy persons, depend on having extended the latter
measurement to the tips of the toes. In other cases the body
and limbs have been measured along the curves instead of
straight from point to point, which naturally has increased the
general dimensions.

The Mayas Kassa, which is the more common species of
orang-utan here, was now becoming well known to me, for I
had in my possession a male and two females quite adult,
besides a young one. The male, as I have remarked before,
differs very slightly from the female. I only noticed a small
difference in the teeth, which may possibly have been acci-
dental. The male has a very small gap between the canines
and incisors, but in the female this space is more marked.

I had heard of two other kinds of orang-utan, one called
Mayas Rambei, the other Mayas Tjaping. The first appeared to
be only slightly different from the Mayas Kassa, being
described as smaller, but with longer hair. The Mayas
Tjaping, however, was very remarkable on account of its
great size and the strange expansions which widen its face. It
appeared also to be much scarcer than the Mayas Kassa, and
I offered a reward of six dollars for every specimen brought
to me in good condition. I also gave special instructions to
the village hunters as to eviscerating the animal, and removing the larger muscular masses in order to lessen the
weight and prevent decomposition, so rapid in this climate;
and this without injuring the skin.

This morning one of my Malays escaped to Simanggan, for
some reason unknown to me; but the Malays are a strange
people, and even their ideas appear to be nomadic, just like
the life they best like. I at once engaged a Dyak in his stead,
a youth named Pagni, who proved also useful in aiding me to
compile a small Sea-Dyak dictionary for my own use—a language much more distinct from the Malay than is that of the
Land-Dyaks.

45

The Dyaks of this part of the country are now quiet, and their devotion to the Tuan Muda may be said to be unbounded. They are at present also on good terms with the Chinese, but I believe not from any love for them, and were it not for fear of the Rajah, many a Chinaman's head would even now be added to the grim trophies hanging over the fireplaces of the Dyak houses. More than once, jokingly of course, when on a visit to me at the Kunsi's house, they asked my permission to cut off the heads of the Chinamen, but I am pretty sure that the joke concealed a covert hope that I might grant them leave.

I had no reason to complain of the Chinese, but they had been grumbling and expressing the wish that I should cease preparing Mayas skins in their house. And, indeed, I must confess that they were not entirely without excuse, for the odour of the skins and skeletons, done in the rough, was not too pleasing, although I sprinkled them abundantly with carbolic acid. The Chinese soon learnt to appreciate the antiseptic virtue of the latter, and every morning one or the other would come and beg me to dress some sore or old wound with carbolic solution.

My orang skins caused me much trouble and anxiety, for the damp, combined with the heat, made it most difficult to dry them properly, and to prevent the cuticle from peeling off and the hair from falling. To add to these difficulties the specimens were all very fat, and it was indeed by no means an easy task to clean the skins thoroughly.

Marop is an excellent station for a zoologist, but a poor one for a botanist. Wherever the Dyaks had not made rice fields, the forest had been long devastated in search of rotangs [rattan], bark, and timber for building houses, etc.; and this had rendered the more useful natural products scarce. I can easily understand how edible wild fruits or plants of economic value can disappear, with the native system of cutting down every tree of such a nature. Nearly the whole extent of country I could see around Marop from the hills was in this condition; or else covered with secondary jungle, which had grown where the primeval forest had been destroyed. This is usually invaded by a large fern (*Pteris arachnoidea*, Kauff.) called rassam by the Malays, which produces long tough stalks, and, being also semi-scandent, so binds together the underwood as to render it practically impenetrable, and where it abounds one is obliged to cut a passage through the jungle with the parang. Large areas of the country are also covered with the common lalang grass, and with thickets of 'onkodok' (the common *Melastoma*). Such are in Borneo the 'bad lands' for the botanist.

The bits of primeval forest which I had noticed on my way up to Marop from the landing place on the river had evidently never been turned into rice fields on account of their sterility, the soil being entirely formed of white crystalline sand. The trees there were small and somewhat stunted, but many species I found to be peculiar and not growing in other places in the neighbourhood. Although formed by different species, I believe that the areas covered by this kind of forest correspond to those of the *mattang* mentioned in previous chapters, and I am disposed to regard them as ancient islands, as it were, left high and dry, on which the vegetation has continued unchanged since the time when they were surrounded by the sea. The 'mattangs' appear to me to have a certain analogy with the 'campos' of Brazil, which might also be considered ancient islands which have been surrounded with alluvial lands of recent formation. This hypothesis would account for the special character of the forest in such localities, so different from that of the country all round.

On returning one day from my daily morning excursion to the forest in search of new plants for my herbarium, I had sat down to skin the baby Mayas brought to me with the first one I had prepared. I had tried to keep it alive, but it had a broken arm, and had been badly shaken, so that my care was of no avail, and it died. Whilst I was thus engaged, Atzon, my best Chinese hunter, came in with a magnificent specimen of the Mayas Tjaping tied to a pole and carried by two men, who, however, had been obliged to get help on the way from the Dyaks, the weight being too much for them. Entire, I do not believe that the creature weighed less than 16 stone. Following my directions, the viscera had been properly extracted without damaging either skin or bone; a large part of the bigger muscles had also been removed, and it was thus in excellent condition. It was also quite fresh, having been killed in the gloaming of the previous evening whilst asleep with its head on its hand on a big branch. It showed only one wound, near the coccyx, the bullet having penetrated all the viscera without touching a single bone.

It was a fully adult male, but the more experienced hunters maintained that it had not by any means attained its full dimensions. Atzon assured me that he had once killed a much bigger one, very old, with hair nearly white, having lost its canine teeth through age. Before skinning the animal my measurements, taken with precautions already mentioned, gave the following results:

Total height from crown to sole of the feet (Some little addition should be made to this measurement, for the body was stiff and the legs much bent)		1.260 m.
Width of the extended arms		2.430 ”
Length of trunk, crown to coccyx . . .		0.915 ”
Circumference of thorax, just below sternum .		1.090 ”
” of neck		0.700 ”
” of forearm		0.350 ”
” of arm		0.330 ”
” of thigh		0.470 ”
” of leg		0.300 ”
Width of the face		0.323 ”
Length of the face		0.310 ”

The face is, beyond doubt, the most singular feature in this animal. Certainly, considering that it is one of the anthropoids, the resemblance to that of Man is very much hidden, I may well say, masked; and it is certainly less human than that of the Mayas Kassa. The flat circular face of the Mayas Tjaping is very much like that of the moon as given in popular almanacks. The eyes are on a level with the skin, somewhat like those of a Chinese, small, and with a chestnut-brown iris, while the very small amount of sclerotic which is exposed at the corners of the eye is very dark in colour.

The singular shape of the face of the Mayas Tjaping is due to the expansions of the cheeks, caused by an accumulation of fat just over the masseter muscles in front of the ears, which are thus hidden from view when the animal is looked at from in front. These expansions are compressed and laminar, about an inch and a half thick, and not rounded as they are reproduced in badly mounted museum specimens. The skin over them is tense and smooth. Except as regard their position, they may be compared to the protuberances on the face of *Sus verrucosus*, or to the hump on the back of

Indian cattle. The colour of the naked portions of the face is nearly black, or, rather, blackish olive. The body is covered with very long hair of a deep fulvous red.

The skin was very thick and tough, and the operation of taking it off extremely arduous and unpleasant, for I had to work on the ground without proper tools, tormented all the time by ants, flies, horse-flies, and mosquitoes, not to mention the excessive heat and the unpleasant emanations. A Chinaman and my Dyak boy Pagni helped me pretty well to get off the fat and clean the skin, and afterwards to take the flesh off the bones.

Whilst I was thus hard at work another Mayas Kassa was brought in, but it had been so badly mauled that neither the skin nor the skeleton were worth preserving, even had I had time to attend to it. It was pregnant, I learnt, but unfortunately the fœtus had been taken out and thrown away with the viscera. I had put the skin of the already mentioned baby orang-utan with a broken arm into spirits, for the huge Mayas Tjaping took up all my time; in fact, I worked at its preparation all that day, all the next, and part of the third. I was obliged to incise longitudinally each of the fingers and toes to clean them thoroughly; even the terminal phalanges were taken out, so that both skin and skeleton should be complete. I dressed the bones well with arsenical soap, which prevented putrefaction, and kept them from the ravages of animals, and, tying them up together in a bundle, I hung them under the roof of a hut which was occasionally used as a blacksmith's shed, where they could dry without giving me further trouble. But the task of preserving the skin was another affair altogether, for the season was rainy and the dampness excessive. I therefore covered it on both sides with arsenical soap, wherever the hair did not prevent it, and, placing it on a bamboo grating, where it lay flat, I hoisted it up under the roof in the middle of the hut, so that it might dry well with plenty of air all round. If necessary, I might have lighted a fire in the hut to dry the air—not to attempt to dry the skin by such means, which would have been a great mistake. Skins of animals collected in tropical climates where the air is damp should never be dried over a fire or exposed to the sun's rays, for by so doing they undergo a sort of cooking, and either get excessively brittle, or else remain liable to absorb damp, so that it is difficult to mount them afterwards as museum specimens, for if they do not fall to pieces they lose both cuticle and hair.

The consequence of this hard work on big mammal skins and skeletons with inefficient tools was that my hands and fingers were more or less cut, and the arsenic getting into the wounds and under the nails caused painful sores, which suppurated.

On the first of April fine weather returned, and we had a bright sun and a pleasant breeze. This was good for my skins, whose preservation was causing me no little anxiety. I had not only to fight against the pernicious effects of the climate, but against ants, rats, and, above all, dogs. Of the latter no less than seven were kept in the Kunsi's house, and fattened to be eaten on grand occasions. Notwithstanding my constant attention, and although I placed the skins in positions which I fancied to be quite secure, I discovered that the heel of one of them, which was nearly dry, had been gnawed. A dog had done the damage, and had got at the skin by climbing up a pole, just like a cat. Certainly, up to that date I had no idea that Chinese dogs were capable of climbing.

For several days I had been aware that the Kunsi was not pleased at my being in his house, and would have been glad to see me go elsewhere. He said that the orang-utans stank and spoilt his meals. This may have been true, although a horror of bad smells is scarcely what one would expect in a Chinaman, but I believe the real fact was that he attributed a malevolent influence to my work, fearing, perhaps, that the irate spirits of the big apes might wander near their mortal remains and clamour for vengeance. I was very nearly obliged to employ violence whilst skinning the big Mayas Tjaping, for the Kunsi wanted it carried out of the house. The Dyaks present grinned, and whispered to me not to bother, and that if I only said the word they would soon have the heads of all those Chinese pigs.

From what I could make out the diabolic influence of my deeds was considered already to be at work, having prostrated an old Chinaman by severe illness; but I believe that the poor fellow was already ailing, and suffering from an attack of typhoid fever when I arrived at Marop. The Chinamen, however, had got it into their heads that my orangs had reduced him to a dying condition. I witnessed the singular treatment to which they subjected the poor sufferer. They made him swallow two pills as big as cherries, of a composition unknown to me, poking them down his throat with their fingers. He was then obliged to smoke opium several times, walking up and down the room, and when he could no longer move through sheer weakness, they put him to bed, taking thither the opium-smoking apparatus. To get him away, I believe, from the evil influences of which I was the cause, they carried him to another house. But as he was in a high fever, they soon after took him down to the stream, and kept him immersed in the water for a quarter of an hour. Apparently the use of a bath to keep down fever has been practised in China long before it was known to us. After the bath they made him swallow two bananas, and then obliged him to smoke opium repeatedly. The next morning the poor old man was dead, which was not surprising. And yet they believed that my Mayas had killed him!

On the 3rd April the weather was again damp and rainy, and I became anxious about my orang skins. I accordingly had a fire lighted in the smithy to endeavour to keep the air in the hut as dry as possible. After breakfast I was told that a Mayas had been seen in the vicinity, so I sallied forth with my gun and followed my guides. In less than twenty minutes they showed me a big tree, about 150 feet high, on which, sure enough, I saw the animal, still in the same place where it had been first seen. It was partially hidden amidst the branches, and would not move, although we made plenty of noise. From where I stood at the foot of the tree it was a difficult shot, for I had to aim nearly vertically upwards. I fired first one and then a second shot, but could not make out whether I had hit him or not, he then slowly moved, but did not leave the tree. This was growing at the bottom of a deep ravine, so I climbed up one of the slopes, and was then able to see the creature well; it was looking down, and was evidently badly wounded. I got a good position, and, after a careful aim, fired again. This time the Mayas fell crashing through the branches, which happily somewhat broke its fall, or, from the immense height of its perch, it would have reached the ground a bag of broken bones. When I got to it, it was quite dead. My last bullet had gone clean through its heart and had passed out at the nape of the neck, splitting the occipital bone. I noticed that as soon as it fell it gave off a peculiar odour of venison. It proved to be a half-grown male, and the girth of the thorax, just below the sternum, was 62 centim. I preserved the skin of this specimen in spirits, and on my return presented it to my former teacher in zoology, Professor Paolo Savi, of the University of Pisa, where it is now mounted in the Zoological Museum.

Wanderings in the Great Forests of Borneo, translated by Enrico H. Giglioli, and revised and edited by F. H. H. Guillemard, Archibald Constable and Co. Ltd., London, 1904, pp. 137–52.

Early European Contacts with Malay Sultanates and Dayak Tribes

4
The Sultanate of Brunei in 1521

ANTONIO PIGAFETTA

Antonio Pigafetta of Vicenza and Knight of Rhodes, an Italian, who accompanied Ferdinand Magellan on the first voyage round the world, wrote the first detailed account of the Kingdom of Brunei, located on the north-west coast of Borneo. Magellan's fleet had sailed from Seville in Spain on 10 August 1519. It journeyed across the Atlantic and Pacific Oceans, reaching the Philippines on 16 March 1521. Magellan died eleven days later in an armed encounter with natives on the island of Matan, near Cebu. The voyage continued and reached Borneo in early July 1521. Eventually, the fleet cast anchor back at Seville on 8 September 1522 after a voyage lasting over three years.

This account, based on a brief sojourn in the port-capital of Brunei, is part of a shortened version of a much larger work which has not survived. It was first published in French in about 1525. In addition to this manuscript, two others in French are also known and one in Italian. As Robert Nicholl has said: 'Pigafetta's account became the basis of all sub-sequent accounts of the visit to Brunei' (1975: 13). The later versions included those by Gonzalo Fernandez Oviedo y Valdés and Francisco Lopez de Gomara.[1]

---

[1]Robert Nicholl (ed.), *European Sources for the History of the Sultanate of Brunei in the Sixteenth Century*, Brunei Museum, Brunei, 1975.

Pigafetta provides evidence of the wealth, power, and cul-
tural sophistication of the Muslim Sultanate of Brunei at a
time when it was in the process of extending its territorial con-
trol along the coasts of Borneo and into the Philippine islands.
In the sixteenth century, especially, Brunei enjoyed a position
of considerable importance in the commercial and political
relations of the South China Sea region.

GOING from Palaoan towards the South-west, after a
run of ten leagues, we reached another island.* Whilst
coasting it, it seemed in a certain manner to go
forward; we coasted it for a distance of fully fifty leagues,
until we found a port. We had hardly reached the port when
the heavens were darkened, and the lights of St Elmo appeared on our masts.

The next day the king of that island sent a prahu [native
boat] to the ships; it was very handsome, with its prow and
stern ornamented with gold; on the bow fluttered a white and
blue flag, with a tuft of peacock's feathers at the top of the
staff; there were in the prahu some people playing on pipes
and drums, and many other persons. Two almadias followed
the prahu; these are fishermen's boats, and a prahu is a kind
of fusta. Eight old men of the chiefs of the island came into
the ships, and sat down upon a carpet on the poop, and presented a painted wooden vase full of betel and areca (fruits
which they constantly chew), with orange and jessamine
flowers, and covered over with a cloth of yellow silk. They
also gave two cages full of fowls, two goats, three vessels full
of wine, distilled from rice, and some bundles of sugar cane.
They did the same to the other ship; and embracing us they
departed. Their rice wine is clear like water, but so strong
that many of our men were intoxicated. They call it arak.

Six days later the king again sent three very ornamented
prahus, which came playing pipes and drums and cymbals,
and going round the ships, their crews saluted us with their
cloth caps, which hardly cover the tops of their heads. We
saluted them, firing the bombards without stones. Then they
made us a present of various victuals, but all made with rice,
either wrapped in leaves in the form of a long cylinder, or in
the shape of a sugar loaf, or in the shape of a cake, with eggs

*Borneo.

and honey. They then said that their king was well pleased that we should make provisions here of wood and water, and that we might traffic at our pleasure with the islanders. Having heard this, seven of us entered one of the prahus, taking with us presents for the king, and for some of his court. The present intended for the king consisted in a Turkish coat of green velvet, a chair of violet coloured velvet, five ells of red cloth, a cap, a gilt goblet, and a vase of glass, with its cover, three packets of paper, and gilt pen and ink case. We took for the queen three ells of yellow cloth, a pair of slippers, ornamented with silver, and a silver case full of pins. For the king's governor or minister three ells of red cloth, a cap, and a gilt goblet; and for the herald who had come in the prahu, a coat of the Turkish fashion, of red and green colours, a cap and a packet of paper. For the other seven chief men who had come with him, we prepared presents; for one cloth, for another a cap, and for each a packet of paper. Having made these preparations, we entered the prahu, and departed.

When we arrived at the city, we were obliged to wait about two hours in the prahu, until there came thither two elephants covered with silk, and twelve men, each of whom carried a porcelain vase covered with silk, for conveying and wrapping up our presents. We mounted the elephants, and those twelve men preceded us, carrying the vases with our presents. We went as far as the house of the governor, who gave us supper with many sorts of viands. There we slept through the night, on mattresses filled with cotton, and covered with silk, with sheets of Cambay stuff.

On the following day we remained doing nothing in the house till midday, and after that we set out for the king's palace. We were again mounted upon the elephants, and the men with the presents preceded us as before. From the governor's house to that of the king, all the streets were full of men armed with swords, spears, and bucklers, the king having so commanded. We entered the palace still mounted upon the elephants; we then dismounted, and ascended a staircase, accompanied by the governor and some of the chief men, and entered a large room full of courtiers, whom we should call the barons of the kingdom; there we sat upon a carpet, and the vases with the presents were placed near us.

At the end of this hall there was another a little higher, but not so large, all hung with silk stuffs, among which were two curtains of brocade hung up, and leaving open two windows which gave light to the room.

There were placed three hundred men of the king's guard with naked daggers in their hands, which they held on their thighs. At the end of this second hall was a great opening, covered with a curtain of brocade, and on this being raised we saw the king sitting at a table, with a little child of his, chewing betel. Behind him there were only women.

Then one of the chief men informed us that we could not speak to the king, but that if we wished to convey anything to him, we were to say it to him, and he would say it to a chief or courtier of higher rank, who would lay it before a brother of the governor, who was in the smaller room, and they by means of a blow pipe placed in a fissure in the wall would communicate our thoughts to a man who was near the king, and from him the king would understand them. He taught us meanwhile to make three obeisances to the king, with hands joined above the head, raising first one then the other foot, and then to kiss the hands to him. This is the royal obeisance.

Then by the mode which had been indicated to us, we gave him to understand that we belonged to the King of Spain, who wished to be in peace with him, and wished for nothing else than to be able to trade with his island. The king caused an answer to be given that he was most pleased that the king of Spain was his friend, and that we could take wood and water in his states, and traffic according to our pleasure. That done we offered the presents, and at each thing which they gave to him, he made a slight inclination with his head. To each of us was then given some brocade, with cloth of gold, and some silk, which they placed upon one of our shoulders, and then took away to take care of them. A collation of cloves and cinnamon was then served to us, and after that the curtains were drawn and the windows closed. All the men who were in the palace had their middles covered with cloth of gold and silk, they carried in their hands daggers with gold hilts, adorned with pearls and precious stones, and they had many rings on their fingers.

We again mounted the elephants, and returned to the house of the governor. Seven men preceded us there, carrying the presents made to us, and when we reached the house they gave to each of us what was for him, putting it on our left shoulder, as had been done in the king's palace. To each of these seven men we gave a pair of knives in recompense for their trouble.

Afterwards there came nine men to the governor's house, sent by the king, with as many large wooden trays, in each of which were ten or twelve china dishes, with the flesh of various animals, such as veal, capons, fowls, peacocks, and others, with various sorts of fish, so that only of flesh there were thirty or thirty-two different viands. We supped on the ground on a palm mat; at each mouthful we drank a little china cup of the size of an egg full of the distilled liquor of rice: we then ate some rice and some things made of sugar, using gold spoons made like ours. In the place in which we passed the two nights there were two candles of white wax always burning, placed on high chandeliers of silver, and two oil lamps with four wicks each. Two men kept watch there to take care of them. The next morning we came upon the same elephants to the sea shore, where there were two prahus ready, in which we were taken back to the ships.

This city is entirely built on foundations in the salt water, except the houses of the king and some of the princes: it contains twenty-five thousand fires or families. The houses are all of wood, placed on great piles to raise them high up. When the tide rises the women go in boats through the city selling provisions and necessaries. In front of the king's house there is a wall made of great bricks, with barbicans like forts, upon which were fifty-six bombards of metal, and six of iron. They fired many shots from them during the two days that we passed in the city.

The king to whom we presented ourselves is a Moor,* and is named Raja Siripada: he is about forty years of age, and is rather corpulent. No one serves him except ladies who are the daughters of the chiefs. No one speaks to him except by means of the blow-pipe as has been described above. He has ten scribes, who write down his affairs on thin bark of trees, and are called *chiritoles.*** He never goes out of his house except to go hunting.

On Monday, the 29th of July, we saw coming towards us

*Muslim
**Writers of narratives.

'A Street in Bruni', *Illustrated London News*, 13 October 1888.

more than a hundred prahus, divided into three squadrons, and as many *tungulis*, which are their smaller kind of boats. At this sight, and fearing treachery, we hurriedly set sail, and left behind an anchor in the sea. Our suspicions increased when we observed that behind us were certain junks which had come the day before. Our first operation was to free ourselves from the junks, against which we fired, capturing four and killing many people: three or four other junks went aground in escaping. In one of those which we captured was a son of the king of the isle of Luzon, who was captain-general of the King of Burné,* and who was coming with the junks from the conquest of a great city named Laoe, situated on a headland of this island opposite Java Major. He had made this expedition and sacked that city because its inhabitants wished rather to obey the King of Java than the Moorish King of Burné. The Moorish king having heard of the ill-treatment by us of his junks, hastened to send to say, by means of one of our men who was on shore to traffic, that those vessels had not come to do any harm to us, but were going to make war against the Gentiles, in proof of which they showed us some of the heads of those they had slain.

Hearing this, we sent to tell the king that if it was so, that he should allow two of our men who were still on shore,

*Borneo or Brunei.

with a son of our pilot, Juan Carvalho, to come to the ships: this son of Carvalho's had been born during his first residence in the country of Brazil: but the king would not consent. Juan Carvalho was thus specially punished, for without communicating the matter to us, in order to obtain a large sum of gold, as we learned later, he had given his liberty to the captain of the junks. If he had detained him, the King Siripada would have given anything to get him back, that captain being exceedingly dreaded by the Gentiles who are most hostile to the Moorish king.

And, with respect to that, it is well to know and understand that in that same port where we were, beyond the city of the Moors of which I have spoken, there is another inhabited by Gentiles, larger than this one, and also built in the salt water. So great is the enmity between the two nations that every day there occurs strife. The king of the Gentiles is as powerful as the king of the Moors, but he is not so proud; and it seems that it would not be so difficult to introduce the Christian religion into his country.

As we could not get back our men, we retained on board sixteen of the chiefs, and three ladies whom we had taken on board the junks, to take them to Spain. We had destined the ladies for the Queen; but Juan Carvalho kept them for himself.

The Moors of Burné go naked like the other islanders. They esteem quicksilver very much, and swallow it. They pretend that it preserves the health of those who are well, and that it cures the sick. They venerate Mahomed and follow his law. They do not eat pig's flesh..... With their right hand they wash their face, but do not wash their teeth with their fingers. They are circumcised like the Jews. They never kill goats or fowls without first speaking to the sun. They cut off the ends of the wings of fowls and the skin under their feet, and then split them in two. They do not eat any animal which has not been killed by themselves.

In this island is produced camphor, a kind of balsam which exudes from between the bark and the wood of the tree. These drops are small as grains of bran. If it is left exposed by degrees it is consumed: here it is called capor. Here is found also cinnamon, ginger, mirabolans, oranges, lemons, sugarcanes, melons, gourds, cucumbers, cabbage, onions. There are also many animals, such as elephants, horses, buffaloes, pigs, goats, fowls, geese, crows, and others.

They say that the King of Burné has two pearls as large as a hen's eggs, and so perfectly round that if placed on a smooth table they cannot be made to stand still. When we took him the presents I made signs to him that I desired to see them, and he said that he would show them to me, but he did not do so. On the following day some of the chief men told me that they had indeed seen them.

The money which the Moors use in this country is of metal, and pierced for stringing together. On one side only it has four signs, which are four letters of the great King of China: they call it *Picis*. For one cathil (a weight equal to two of our pounds) of quicksilver they gave us six porcelain dishes, for a cathil of metal they gave one small porcelain vase, and a large vase for three knives. For a hand of paper they gave one hundred picis. A *bahar* of wax (which is two hundred and three cathils) for one hundred and sixty cathils of bronze: for eighty cathils a bahar of salt: for forty cathils a bahar of *anime*, a gum which they use to caulk ships, for in these countries they have no pitch. Twenty tabil make a cathil. The merchandise which is most esteemed here is bronze, quicksilver, cinnabar, glass, woollen stuffs, linens; but above all they esteem iron and spectacles.

Since I saw such use made of porcelain, I got some information respecting it, and I learned that it is made with a kind of very white earth, which is left underground for fully fifty years to refine it, so that they are in the habit of saying that the father buries it for his son. It is said that if poison is put into a vessel of fine porcelain it breaks immediately.

The junks mentioned several times above are their largest vessels, and they are constructed in this manner. The lower part of the ships and the sides to a height of two spans above water-line are built of planks joined together with wooden bolts, as they are well enough put together. The upper works are made of very large canes for a counterpoise. One of these junks carries as much cargo as our ships. The masts are of bamboo, and the sails of barks of trees. This island is so large that to sail round it with a prahu would require three months. It is in 5° 15' north latitude and 176° 40' of longitude from the line of demarcation.

On leaving this island we returned backwards to look for a convenient place for caulking our ships, which were leaking, and one of them, through the negligence of the pilot, struck on a shoal near an island named Bibalon; but, by the help of God, we got her off. We also ran another great danger, for a sailor, in snuffing a candle, threw the lighted wick into a chest of gunpowder; but he was so quick in picking it out that the powder did not catch fire.

On our way we saw four prahus. We took one laden with cocoanuts on its way to Burné; but the crew escaped to a small island, and the other three prahus escaped behind some other small islands.

The First Voyage Round the World by Magellan, translated from the accounts of Pigafetta and other contemporary writers. Accompanied by original documents, with Notes and an Introduction by Lord Stanley of Alderley. Hakluyt Society, London, first series No. 52, 1874, pp. 110–18.

5
The Sultanate of Banjarmasin in 1714

CAPTAIN DANIEL BEECKMAN

On 12 October 1713 the East India Company ship, the Eagle Galley, *set sail from England on a commercial venture to south-eastern Borneo. Daniel Beeckman and J. Beacher were in command. The purpose of the voyage was to reopen English trade in pepper and other tropical goods with the Sultanate of Banjarmasin, and to establish whether this port might serve as a staging-post for the British India–China trade. A few years before, an Anglo-Banjarese conflict had resulted in the physical expulsion by the Sultan of the English settlers in Banjarmasin.*

Beeckman reached Dutch Batavia (Jakarta) on the north coast of Java on 20 April 1714, having sailed via the Cape of Good Hope. His ship anchored at Banjarmasin on the south coast of Borneo on 29 June, departing again in late December. Eventually, Beeckman arrived back in England on 29 October 1715, after a voyage of just over two years duration.

In a recent reprint of Beeckman's book, Chin Yoong Fong remarks, in his Introduction, that Beeckman gives us 'a lucid

*and interesting account of the common people of Ban-
jarmasin, their way of life, customs and beliefs' (1973).[1]
Beeckman saw his book as a succinct, factual statement about
this region of the island, which would serve to provide a guide
or handbook on the commerce, land, and peoples of southern
Borneo for those Englishmen who might follow him.*

*This extract, which is part of Chapter II, describes in an
introductory way the island and two main groups in the
south: the coastal Muslim Banjareens (Banjarese) and the
interior pagan Byajos (Biajus/Ngajus). Beeckman also intro-
duces us to the orang-utans or 'men of the forests' as 'creatures of fable' and he tells us of the capital of the Sultanate.
From the later sixteenth century, on account of pepper cultivation and the export of pepper, Banjarmasin became an
important regional power in southern and eastern Borneo.*

THE Island* of Borneo (so called from a City of that
Name) lies on the North of Java, and on the East of
Sumatra, and of the Peninsula of Malacca. It is situate
between the 7 Deg. 30 Min. of North Latitude, and the 4 Deg.
10 Min. of South, under the Equinoctial, which divides it into
two unequal Parts, 7 Deg. 30 Min. lying Northward of it, and
4 Deg.10 Min. Southward: So that it is in Length 700 Miles, in
Breadth 480, and in Circuit about 2000. It is counted the
biggest Island, not only in the Indian Sea, but in the whole
World, except perhaps California in the South Sea.

The Air (considering the Climate) all round the Island
along the Sea Coast is pretty temperate, because of the
refreshing Sea Breezes that blow always about eleven in the
Morning on the South Parts, otherwise the Heat would be
insupportable; but it is very unwholsome because of the
Moistness, in the South Parts especially. For about the River
of Banjar Masseen [Banjarmasin], many score Miles near the
Sea the Country looks like a Forest, being full of prodigious
tall Trees, between which is nothing but vast swamps of Mud.
At high Water you may sail in a great way among these Trees
in several places, but at low Water it is all Mud, upon which

[1]Chin Yoong Fong, 'Introduction' to Captain Daniel Beeckman, *A Voyage
to and from the Island of Borneo, in the East Indies,* Dawsons of Pall Mall
reprint, Folkestone and London, 1973.

*The Old English 'f' in the original text has been converted to modern
English 's' in the remainder of the passage.

the Sun (especially in the Equinox) darting his scorching Beams perpendicularly, raises noisome Vapours, Fogs, etc. which afterwards turn into most violent Showers, that fall more like Cataracts than Rain, and are very cold, being followed generally by cooling Winds; so that the Weather changing suddenly from scorching Heat to chilling Cold, causes the Air to be sickly and unhealthful. In the beginning of the rainy Season there is no sleeping for the Noise which the Frogs make, whereof there is a vast multitude in these swampy Woods: And a great number being left, with their Spawn and other Slime and Filth on the Mud, when the dry Season begins (which is commonly in April, and holds till September) they die, and the Carcasses lie rotting, and occasion a very noisome Stink and Corruption in the Air. During all this dry Season the Wind is Easterly between the South Coasts of Borneo and the Isle of Java; and this is by much the more healthy Part of the Year; but from September, or thereabouts, to about April, the Westerly Winds reign with violent Storms, prodigious Rain, Thunder and Lightning almost daily; for during this Season it is rare to have two Hours of fair Weather in twenty four on the South Coast of this Island; and tho' the other Season is so fair, yet you are sure to have a Shower for about an Hour every Day at the coming in of the Sea Breezes, which cools the Air, and makes it very agreeable.

The Country abounds with Pepper, the best Dragons-blood, Bezoar, most excellent Camphire,* Pine Apples, Pumblenoses, Citrons, Oranges, Lemons, Water Melons, Musk Melons, Plantons, Bonano's, Coconuts, and with all sorts of Fruit that is generally found in any Part of the East-Indies. The Mountains yield Diamonds, Gold, Tin, and Iron; the Forests, Honey, Cotton, Deer, Goats, Buffalo's and wild Oxen, wild Hogs, small Horses, Bears, Tygers, Elephants, and a multitude of Monkeys. Here are small Hog-deers (the Feet of which are often used for Tobacco-stoppers, when tip'd) which they catch in this manner. When they find the track of these Creatures, they dig square Holes in the Earth, about five Foot over and four Foot deep, which they cover over with a little Straw, or such like, and sift some Dust thereon, so that the Hog-deer in passing over falls in. The Monkeys, Apes,

*Camphor.

and Baboons are of many different Sorts and Shapes; but the most remarkable are those they call Oran-ootans, which in their Language signifies Men of the Woods: These grow up to be six Foot high; they walk upright, have longer Arms than Men, tolerable good Faces (handsomer I am sure than some Hottentots that I have seen) large Teeth, no Tails nor Hair, but on those Parts where it grows on humane Bodies; they are very nimble footed and mighty strong; they throw great Stones, Sticks, and Billets at those Persons that offend them. The Natives do really believe that these were formerly Men, but Metamorphosed into Beasts for their Blasphemy. They told me many strange Stories of them, too tedious to be inserted here. I bought one out of curiosity, for six Spanish Dollars; it lived with me seven Months, but then died of a Flux; he was too young to show me many Pranks, therefore I shall only tell you that he was a great Thief and loved strong Liquors; for if our Backs were turned, he would be at the Punch-bowl, and very often would open the Brandy Case, take out a Bottle, drink plentifully, and put it very carefully into its place again. He slept lying along in a humane Posture with one Hand under his Head. He could not swim, but I know not whether he might not be capable of being taught. If at any time I was angry with him, he would sigh, sob, and cry, till he found that I was reconciled to him; and tho' he was but about twelve Months old when he died, yet he was stronger than any Man in the Ship.

As to the Birds, I met with none such as we have in England, except the Sparrow. Here are Parrots and Parrokets of various sorts and sizes, from the bigness of a Bulfinch to that of a Raven; particularly a sort called by the Banjareens* Luree (that are brought hither by the Maccassars**), which they so much admire for their Beauty, Docility and sweet Smell, that there are few Houses without one of them; they give sometimes six or seven pieces of Eight for one; I bought several, but the cold Weather at Sea killed them. Here are such vast multitudes of Bats, that at particular times (*viz.* just before the setting in of the Westerly Monsoon) towards Evening I have seen the Sky almost darken'd by them, when at Tatas, flying

*Banjarese.
**Makassars, from Sulawesi.

From Captain Daniel Beeckman, *A Voyage to and from the Island of Borneo, in the East Indies*, T. Warner and J. Batley, London, 1718.

from the West towards the East for the space of two Hours. I shot one in the Woods, whose Body in Shape, Colour and Smell was like a Fox, having Head, Ears and Teeth, etc. as big as a young one: The Wings when spread, measured from the tip of the one to the tip of the other, 5 Foot 4 Inches.

The Rivers and the Sea Coasts afford plenty of Fish, as Mullets, Breams, etc. a sort of Fish called Cockup, the best tasted foreign Fish I ever met with; and many other sorts which we have not in Europe, particularly the Cat-fish, which is much esteemed by the Natives, but seldom eaten by the English. I think the Flesh of the young ones is of a tolerable relish, but very luscious; there are some of five or six Foot long, they have no Scales, their Heads are large, not unlike a Cat's Head, having Barbs very like a Cat's Whiskers. The River Banjar dischargeth its Waters into the Sea, in the Latitude of 3 Deg. 18 Min. South. It is remarkable, that at the latter end of the dry Season, when the Springs are low, the Water is of a Brackish taste up as high as China River; at which time the Cat-fish follow the Boats in great numbers, and getting under the bottom of them make a dreadful groaning; it surprised me much at first. In this River are caught Prauns, generally six or eight Inches long, also very large Rock Oysters at a little Island called Pooloobatoo [Pulau Batu].

The Natives are of two sorts, *viz.* those that inhabit in or near the Ports of Trade (as particularly the Banjareens) and the Inhabitants of the inland Country; for the former are of a middle Stature, rather under than over, well shap'd and clean limb'd, being generally better featur'd than the Guinea Negroes: Their Hair is long and black, their Complexion somewhat darker than Mulattos, but not quite so black as the aforesaid Negroes; they are affronted if you call them black Men. Both Men and Women value themselves in a particular manner, if they are whiter than ordinary. They are very weak of Body, which is occasion'd chiefly by their lazy unactive Life, and mean Diet not having the opportunity of Walking, or of any Land Exercise, and working seldom, but are always in a sitting posture, either in their Boats or Houses; neither do they stir without it be out of absolute necessity. They us'd to laugh at us for walking about in their Houses, telling us that it looked as if we were mad, or knew not what we did: If, say they, you have any Business at the other end of the Room, why do you not stay there; if not, why do you go thither; why always stalking backwards and forwards? If the Banjareens have but a quantity of Rice and Salt, they think themselves very rich, for if they throw a casting Net at their Door, they need not fear the want of a Dinner, so great abundance of Fish is in that River.

The Women are very little, but very well shaped, having much handsomer Features and better Complexion than the Men; they walk very upright, and tread well, turning their Toes out, which is contrary to the Purchase of most Indians. I believe it is a Custom forced upon them by their walking on the Logs that float upon the River before their Doors from House to House, as I shall explain more at large by and by. They are very constant when married, but very loose when single; neither is her former Compliancé counted a Fault in a Wife; and the Mothers do often prostitute their Daughters at eight or nine Years of Age for a small lucre. They generally marry at that Age, and sometimes under; but as they are soon ripe for Matrimony, their Fertility soon decays, for they are generally past Child-bearing at 20 or 25; it is rare that a Woman holds till 30. They live to a tolerable good Age, and use daily Bathing in the Rivers, and are expert Swimmers. Every Day whilst we remained at Tatas, we saw the River full of Men, Women and Children, even some in Arms, which they carry in for Health's sake, to which this way of Bathing must needs be very beneficial and refreshing in so hot a Climate.

In burying their Dead they take care to lay their Heads towards the North, and put into the Grave with them a great deal of Camphire, and several things necessary for the support of Life; for what end the Camphire is deposited there I know not; but the latter is according to an old Pagan Custom, that has been handed down to them, as believing that those Provisions were useful to them in their Journey to the other World. But now being Mahometans* they say they do it only as a mark of Respect. They carry them in Boats as near as they can to the Burying-place, attended by their Friends in great Order and Ceremony, being dress'd all in White, with lighted Torches in their Hands, tho' it be in the Day-time.

The inland Inhabitants are much taller and stronger bodied Men than the Banjareens, fierce, warlike and barbarous. They

*Muslims.

are called Byajo's,* an idle sort of People, hating Industry or Trade, and living generally upon Rapine and the Spoil of their Neighbours; their Religion is Paganism, and their Language different from that spoken by the Banjareens. They go naked and only have a small piece of Cloth that covers their private Parts; they stain their Bodies with blue, and have a very odd Custom of making Holes in the soft part of their Ears when young, into which they thrust large Plugs, and by continual pulling down these Plugs the Holes grow in time so large, that when they come to Man's Estate, their Ears hang down to their very Shoulders. The biggest end of the Plug is as broad as a Crown piece, and is tipt with a thin Plate of wrought Gold. The Men of Quality do generally pull out their fore Teeth and put Gold ones in their room. They sometimes wear, by way of Ornament, Rows of Tygers Teeth strung and hung round their Necks and Bodies. Those of them that were subject to the Sultan of Caitangee** (whom I shall have occasion to mention often hereafter) are now in Rebellion against him; he that headed them made Pretences to the Crown, and was set up by these Mountaineers against the present Sultan (to whose Government they are very averse) who was chosen by the general Consent of the People, at least of the civiliz'd trading part of them. But this Pretender, before I came away, was dispatch'd by Poison. However, some of those People, *viz.* those that live near the Ports of Trade, are in subjection to their different Kings or Sultans; the others live in Clans by themselves, without Kings, or any Form of Government. I have seen some of the former come down the River to the Port of Banjar Masseen in very ill-shap'd Praws; and bring down Gold Dust, Diamonds, Bezoar-stones, Rattans, and sundry other Merchandizes. The Banjareens will not suffer the Europeans to have any Acquaintance or Trade with them, but do purchase the Goods from them, which they sell to us at a greater Price. And I do verily believe, that the many frightful Stories they tell of those People's Barbarity and Cruelty, are only invented on purpose to deter us from having any Acquaintance or Commerce with them, which would be a great disadvantage to the latter; tho' some of these Reports may be true: As to their Women I never saw

*Biajus or Ngajus.
**Kayu Tangi, near Martapura.

any of them, and so can give no Account of them. The Island is divided into different Kingdoms, having their particular Kings or Sultans, whom they call Raja's.

There are in this Island four chief Ports of Trade, *viz.* the City of Borneo,* situate on the North, in the Latitude of 4 Deg. 30 Min. North. Passeer [Pasir] on the East side, in the Latitude of 1 Deg. 15 Min. South. Succadana [Sukadana] on the West, in the Latitude of 0 Deg. 15 Min. South. And the Port of Banjar Masseen, on the South, in the Latitude of 3 Deg. 18. Min. South. Here was formerly a Town called Banjar, about 12 English Miles from the Sea, built partly upon Floats of Timber, partly upon Stilts; it was near it the English Factory was established; but there is not so much as the Remains of a Town to be seen now, the Inhabitants having removed to other Places, but most to Tartas or Tatas, a City about six Miles further up the River. As to the three former I can give no particular Account of them from my own Observation, but by what I learnt from the Banjareens. As to the last I shall be very particular, all that I shall mention touching it being of my own knowledge, and have taken more Pains than ordinary that I might be more capable of informing the honourable East-India Company of the Methods that may be used in order to settle a Trade there: And I dare say, no Person ever had a greater opportunity of knowing those Matters than myself. I shall only say that there are several Kings or Raja's in the inland Country; as also the Cities of Borneo, Succadana and Passeer have each of them one; that formerly all the other Raja's (as well as he to whom Banjar Masseen belongs) were subject to the Raja of Borneo, who was a Supreme King over the whole Island; but now his Authority is mightily decreased, and there are other Kings equal, if not more powerful than himself, particularly the Sultan of Caitangee. His Name is Pannomboang, and stiles himself Sultan of Caitangee, which is the City where he resides, situate within 100 Miles of the Port of Banjar Masseen. His Brother is another King, and stiles himself Sultan of Negarree,** a City about 300 Miles up the main River, where he resides. But the former is the greatest, by reason of the Trade and the Customs he receives from this Port, which may be computed

*Brunei.
**Negari/Negara.

to amount to 6 or 8000 Pieces of Eight per annum. But I think I have said enough of these general Matters, and it is time to give an Account of our particular Proceedings after our arrival in the River.

After we had cast Anchor as aforesaid, we espy'd a small Praw or Boat under the Shore; we sent in a very civil manner to the Persons that were in it, and intreated them to come on board. We lay then with our English Colours flying, at which they were much surprized, knowing how severely they had used our Countrymen, when last among them. However, partly thro' Fear, and partly thro' our kind Invitation, they came on board. They were very poor-look'd Creatures, that had been at Tomberneo [Tabanio], and were returning to Tatas. We express'd all the Civility imaginable towards them, gave them some small Presents, and desired that they would acquaint their King or Grandees in the Country, that there were two English Ships come to buy Pepper of them; that we were not come to quarrel, but to trade peaceably, and would pay them very honestly, and comply with all reasonable Demands according to what should be hereafter agreed on. They inquir'd whether we were Company Ships, to which we did not readily answer them; but before we did, they proceeded and said, that if we were, they, as Friends, would advise us to depart the Port forthwith, because their Sultan and their Oran-Cays, or great Men, would by no means have any Dealings with us. We design'd to have sent our Boat that Night to their Town call'd Tatas (which is about 30 Miles above the Place where we lay) that she might arrive there by Day-light the next Morning; but those Persons dissuaded us from it, assuring us that we should soon have News from their Sultan; and that some of their Men would not fail to be down with us the next Day. Then they took their leave of us, returning us many Thanks for our Presents.

The next Day came on board of us a Boat, with one Cay Rouden T'acka, and Cay Chetra Uday, being Messengers from the King. We received them as civilly as possible. The first thing they inquired, was whether we were Company Ships, or separate Traders; that if the former, we need not wait for an Answer, and that it would be our best ways to be gone; desiring earnestly that what Answer we should return them might be sincere; for that whatever we said to them should be told the Sultan. Finding no other method to introduce ourselves, we were forc'd to assure them that we were private Traders, and came thither on our own Account to buy Pepper. This we did, believing we might in time have a better Opportunity of making our honourable Masters known, and of excusing the heavy Crimes laid on their former Servants, whose ill Conduct had been the Cause of the Factory's being destroyed. They ask'd us why we came thither rather than to any other Place, since our Countrymen had so grosly abused them? We answer'd, that we were Strangers to that Affair; and that at first we design'd to go to Pallambam;* but being inform'd that Pepper was much cheaper here, we were willing first to try this Market. They also enquir'd what number of Men and Guns we had, and cast their Eyes slyly about to endeavour to guess of what strength we were; for they are exceeding jealous of all Europeans.

Towards Night they departed, and we gave them some Guns. They left two Persons on board, with whom they desir'd that our Linguist would come up to the Town the next Day, to give answer to such other Questions as might be ask'd. We gave Instructions to our Linguist to tell them, that we were two small separate Stock-Ships; that we were inform'd at Batavia, that Pepper was very cheap at this Port, so chose rather to come hither than to Pallambam: We order'd him to learn on what Conditions they would offer to trade with us, and who were the properest Persons to apply to; to press a speedy Meeting: And if they ask'd what we had to purchase Pepper with, to tell them Mexico Pieces of Eight; (for the Pillar-dollars they will not take); to give them kind Invitations to come on board; to write down all Questions and Answers: And if any thing of Consequence should be further ask'd, to give no Assurances or Answers of themselves, but to plead Ignorance, and to refer all to the Merchants; (for so they were to call us, and not Supercargo's, which would have created a Jealousy that we belong'd to the Company;) to take care to keep the Sailors sober, and in good Order; with some other Instructions less material.

Having given them these general Directions, we sent them away the 2d of July at two a Clock in the Morning. One of the Linguists was an Englishman, the other a Javan whom we hir'd at Batavia; but we put most Confidence in the first. They

*Palembang, in Sumatra.

71

return'd that very Night, and told us, we should have an Answer in seven Days from the Sultan of Caytangee, and in eleven from the Sultan of Negarree. They also brought us a Caution from the Banjareens to beware of some large Pirate Praws mann'd with about a hundred of the Byajo Men, that lay skulking thereabouts. But before this Advice came, we were like to have felt some of the cruel Effects of their Barbarity through our own Inadvertency: For that Day about Noon we saw three large Praws under the Shore, which had shot up the River a little above our Ships: Whereupon, imagining they were Banjareens, and hoping to get some better Information in relation to Trade, I went into the Long-boat in Company with Mr Bartholomew Swartz, chief Super-cargo of Borneo, and Mr John Beacher, chief Supercargo of my Ship, and Mr John Gerard our Assistant and Purser, with five Men and a Boy. We carried only two Muskets, and a small Fowling-piece, with two Cartouch-boxes; but had we thought of meeting with such Barbarians as we did, we should have been much better provided. We hoisted our Sail, and stood towards them; but they row'd with all their Might from us; and finding we were like to come up with them, they ran their three Vessels up a Creek among the Trees, which were exceeding thick, hanging over the Water, and gave so great a Shelter that there was no Wind for us to sail up the Creek after them: However, we made in, thinking they were bound no further, But being come close to the Mouth of the Creek, we saw their Praws a little way up, and no men in them; For they, being about a hundred in number, were got ashore among the Trees, designing to draw us in, and destroy us all; which they might easily have done, had they all equal'd the Courage and Resolution of their Leader: For the Creek was not above ten Yards over, and they exceeded us in number above ten to one, being arm'd with Javelins, Sampits, and poison'd Arrows. We call'd aloud, and ask'd them what they had to sell, with some other Questions, but receiv'd no Answer 'till we were got up into the Creek; when on a sudden we heard a horrible Shout, after the manner of these Barbarians; and at the same time their Captain advanc'd boldly towards the Boat, threw a Javelin at us, and immedi-ately after shot an Arrow. It was fortunate for us that his Men were not so forward, and seem'd dismayed, keeping back among the Trees, but let fly a Shower of their poison'd

Arrows among us, which however did us no Damage. We immediately put ourselves in a posture of Defence, and presented our small Arms, but were at first unwilling to fire, least such a Proceeding should frustrate our Design of trading in the Port. But seeing no other Remedy, and perceiving by their Dress and Language that they were not Banjareens, we discharged our Pieces at them, which put them to flight, scouring in among the Trees; tho' even in their Retreat they ceas'd not to let fly their Arrows at us, after the manner of the ancient Parthians. Whilst Mr Gerrard was loading our Guns again we us'd our Pocket-pistols, firing wherever we saw a Bush wag. In the mean time the Sailors were in great Confusion, but not idle, haling the Boat by the means of the Boughs and Shrubs, until they got her out; before which we had discharged our Pieces a second time. But we saw no more of these Villains, they being frightened at the Noise, and Danger of our Fire-arms. We were not a little pleased at our narrow Escape. What loss the Enemy had we know not; but our good Fortune brought us off without so much as one Wound. We brought away some Darts that stuck in the side, and Sail of our Boat. These People go naked, having only a Chawat, or small Piece of Cloath, about the breadth of a Hand, to cover their Privy-parts. Their Bodies were all over stain'd with Blue; and they seem'd to be strong, tall Men, like the Mountaineers spoken of before.

We remain'd on board without any further Answer 'till the 6th of July, when Cay Rouden Taka, and Cay Chetra-Uda came on board. They brought us Presents of Fruit, Fowls, Eggs, etc. which we had rather been without, knowing that our Return for those Trifles must be expensive, tho' we were too often troubled with such mercenary Civilities.

They told us, that the Sultan's Pleasure was, that we should come up to Tatas to hold a Bechara, or Consultation, with Pangarang Purba Negarree, or Prince of Negarree, (who is a Prince of the Royal Blood) promising, that they would stay as a Pledge 'till our Return. Whereupon the other Supercargo's went away on the 7th of July about 11 in the Morning, Captain Lewis, and my self, remaining on board to entertain those Grandees. When they arrived at Tatas, they were introduc'd to the Prince; who, upon Enquiry, understanding that they were not the Captains, he order'd them to bring us along with them to the next Consultation. He examin'd them strictly whether we were separate Traders; which being affirm'd, he said, That we were welcome; and asking what Quantity of Pepper would load us, he was answer'd 4 or 5000 Peculls. We had presented him with an extraordinary good Silver Watch, and in return he promis'd mighty Services; but at the same time told them, that the Sultan being now in War with the Rebels, who inhabit the Pepper Country, that Commodity was grown very dear; therefore he cou'd not come to an exact Price, nor sign a Contract, because the Sultan's Great Seal was not there, with which all such Contracts are sign'd. It is remarkable, that no Business can be done in those Parts, nor scarce Admittance gain'd to any of their great Men, 'till he understands by his Servants that the Person is not come empty handed.

The next Evening the Supercargo's came back, and on the 11th the aforemention'd Cays return'd on board, and desir'd us to go with them to the Prince. They staid with us that Night, and we made them very merry; for though it be against their Religion to drink strong Liquors, yet we soon perceived that they were no Enemies to Arrack or Wine. The next Day I set out about two in the Morning with the other Supercargo's; some in mine, and the rest in the *Borneo*'s Pinnace; having order'd my two Trumpeters to attend us, that we might appear with more advantage.

We went up the River for about 22 Miles, where we turn'd off into a narrow Branch of it. This River is extreme pleasant, being about twice as broad as the Thames at Gravesend, having a vast number of prodigious tall Trees on each side that are always green. I remark'd here four very agreeable Islands at some Miles distance one from another, each being situate about the middle of the River. The first is called Pooloococket [Pulau Kaget], being cover'd with Trees, some of which are of a vast height. You may see it before you enter the River, and it serves as a Land-mark to sail over the Bar. There is a large Sand spits out all round it, but shoots itself out farther at the North and South-end; which must be carefully avoided; for if a Ship ground, the Ebbs are so very strong, because of the Land-waters, that it might wring her to Pieces; besides, great Drifts of Trees come down the River continually, which in such a case would be of ill consequence. Besides, it is somewhat dangerous, because of the often shifting of the Sand: The best Advice I can give is to anchor any where about a Mile or two within the River's Mouth, where the Ground is clear, and Water enough from side to side: Then send your Boat to sound off the aforesaid Island, and buoy it; which after you have pass'd, keep the Starboard Shore on board within your Ship's length; or, if you please, nearer, at any time of Tide, but best on the Flood; for the Ebbs run so strong in some Seasons, that for want of Wind, which the Trees keep off, you'll find it very difficult to get a head when you come to open the first River, on your Starboard side, which is pretty broad, and is call'd China-River as you go up. You must sheer off towards the middle, to avoid a Spit that shoots out from the Larboard Entrance of that River's Mouth; but you have gradual Soundings.

The Tide flows here but once in 24 Hours, and that always in the Day-time; in the Spring-tides the Water rises about 12 Foot; but in the Night there is only a kind of Stagnation of the Water when the Tide comes in, and never rises above half a foot, unless it be in very dry Weather: The Reason is, that besides the strong Current of the Inland Waters, the Land-winds, which blow always in the Night with more Vigour than at other times, make so great a Resistance that the Flood cannot rise to any considerable height.

This River China runs up as far as the Town of Tatas, and is navigable not only thither, but a considerable way farther for the biggest Ship in the World. All the China Junks go up the said River, and from hence I suppose it has its Name. All other Ships that will go up so far must take the same Course; but our Ships lay higher up the main River, over-against the Factory, which was at the Entrance of another River, smaller than the former, and which you meet next on the Starboard side also as you go up: It is called Tatas-small-River; between which place, if you keep nearest the Starboard side, there is no Danger. We pass'd China River about nine or ten in the Morning, and about eleven got to the small River last mention'd, which is much the nearest way for small Vessels or Boats. It was no little Diversion to us in our Passage, to see the prodigious Multitude of Monkeys and Baboons of all sorts that swarm'd on the Trees on each side of us; several with their young ones hanging about their Necks. We shot many of them, but at the report of a Gun they make a terrible bustle with their jumping and scouring from Tree to Tree: They would shake off their young ones, and make the

Woods ring and echo with their loud squealing: The Sound of our Trumpets had the same effect on them. The Natives never hurt them, which makes them so void of fear, that they will let you come very near them. We saw many Alligators sunning themselves on the Mud, several of which we shot at, but to no purpose.

We had eight Miles to go up this River, which is very crooked, where the scorching Heat of the Sun would have been as troublesome to us as these Sights were diverting; but it being narrow, and the Trees wonderful high on each side, we were pretty well shaded from the Heat. We could not see the Town 'till we were just entring into it, because of the tall-ness of the Trees that stand close together. It consists of about three hundred Houses, most of them built on Floats in the River, which is here about a hundred Yards over; but the Houses of the poorer sort are built on Stilts in the Mud on each side. The Owners are forced at high Water to make use of Boats to get into the Houses; and at low Water they have large Logs that lie from House to House, on which they walk. The Houses on the Floats are built on vast Logs or Trees laid and trunnel'd together, or bound strongly with Cables made of Rattans, and fasten'd by the like Cables to the Trees on shore, and to one another. Each House consists only of one Floor, divided into sundry Apartments, according to the Family; the sides being only split Bamboo plated cross-wise; and they are thatched on the top with Cajans, much after the manner of the Javans, and other Malayans. Tho' these Houses are tolerably high for the sake of Air, yet Eaves hang over the sides within five Foot of the Logs or Stage they are built upon, to keep out the Sun. Here runs a very strong Ebb, which sometimes breaks their Moorings or what fastens them to the shore; and you may see three or four Houses adrift at a time. I have been inform'd, that some Houses having broke loose in the Night, whilst the People were asleep, drove out into the main River, and thence to Sea; which is very prob-able, because the Flood sometimes runs very weak, when, on the contrary, the Ebb is exceeding rapid and strong.

When I first got into the Town, I was surpriz'd to see these floating Houses, and the People in great Numbers paddling up and down from House to House in small, but neat built Canoes or Praws. The Curiosity of seeing us had brought a great many to Town from all Parts of the Country, which caused it to be more crowded with People and Boats, than it had been in many Years before. On our first Entrance I order'd the Trumperts to sound; the fine Echo from the Woods and Waters added to the Harmony of our Musick. Most of the Natives were astonish'd at the sudden Noise, and some fled one way and some another in their little Boats, with all the Confusion of a frighted Multitude. The number of ugly black Ferry-men was so great, the stink of the Oil or Ointment, wherewith they besmear their Bodies daily so noisome, and the sultry Heat so excessive, that I had almost persuaded my self I was passing the River Styx, or Phlegeton in Hell. We were conducted to Cay Arrea's House, who is the principal Trader in town, but not a very strict Observer of Justice, and was afterwards introduc'd into the Prince's Presence, who sate cross-leg'd at the upper end of the Room, with Cay Arrea on his left, and Cay Demon on his right. There was also the Chief of the Chinese, who lives there, and is a very considerable Trader, besides several other great Men. We were order'd to sit down cross-leg'd, just opposite to the Prince; which we no sooner did but the House was immediately fill'd with other Indians of the meaner sort, who sate down behind us; so that we were almost stifled with Heat, and the Stench proceeding from their abundant Perspiration. Our Crew waited with the Pinnaces at the Door.

The Prince, with a very reserv'd Countenance, after a profound Silence, spoke first to us, and let us know that they had great Reason to be jealous of all Europeans; and that the Sultan did insist that we should bring up our Ships into the narrow River, or even into the Town, as a Security for the Safety of his Subjects; that it would forward our Loading, and be other ways advantagious. We excus'd it in the handsomest manner we could, telling him that our Men being us'd to a colder Climate, could not live in that warm Situation without the Sea-breezes. This indeed was one Reason; but the chief was, that if we should comply, we must be subject to their Power to use us as they pleased, should they at any time discover that we were Company Ships, by way of Reprizal, and in revenge for the Injuries they complain'd to have been done by our former Factory; and knowing that they were as willing to take our Money as we their Pepper, we absolutely refus'd to yield to that Proposal; and told him, that we would return to our Ships, and stay three Days for an Answer, but no longer. We discours'd on divers Subjects for about three Hours: And when we were about to depart, the Prince desir'd our longer stay, because the Sultan had given strict Orders to Cay Arrea to entertain us whenever we came very handsomly at his Charge. And immediately several large Gold and Silver Bowls neatly wrought were brought in full of Rice, boil'd Fowls, hard Eggs, etc. We eat plentifully, and drank our own Wine and Punch; their best Liquor being the River Water that runs before their Doors. After we had done, what was left was given to our Boats Crew in Brass Bowls. The Prince, while we were at Dinner, withdrew, and din'd by himself; after which he came in again. He was dress'd, after their manner, in Scarlet and Blue, having on a small close-bodied Wastcoat, without a Shirt, and over that a Chawat [loincloth], wrap'd round once or twice, that hung down to his Knees; he wore Drawers, but his Hands, Legs and Feet were bare. On his left side, in a neat Belt stuck a Creice or Dagger, richly set with Diamonds; before him was a Table about two Foot long, and one and a half broad of solid Gold, much like a hand Tea-Table; on which always stood his Furniture for his Betle-nuts, Seree Leaves, and Lime, which he chews continually; as it is the Custom for Men, Women and Children to do in that Country, and to smoak Tobacco. The Box that held the Nuts was not unlike a Rummer, with a Cover to it; that for the Leaves like a standing Snuffer-case; and that for the Lime was a small, round, flat Box: All of the finest Gold, very neatly wrought in Filigreen, and set with large Stones, some Diamonds, and others that I knew not.

Having resolved not to bring our Ships up, nor to stay longer than the time we mentioned, we took our leaves about four in the Afternoon, and the same Night we arriv'd on board, being extremely fatigu'd. On the 13th we sail'd up the River about twenty Miles, and anchor'd over-against the Mouth of China River.

On the 16th came on board the same Messengers from the Prince, and signified, that he desir'd to speak with us again. We feign'd an Indifferency, and told them, that we were then ready to depart, since we cou'd not agree. We soon perceiv'd that this News did not please them. However, we told 'em, we would go up once more; and accordingly next Day early in the Morning we set out, and arrived there about two in the Afternoon. We were again introduc'd to the Prince; and after several Hours Discourse, we over-rul'd the Proposal of bringing our Ships up to Town; and only complied with his Request of taking a House in Town, where the Supercargo's should reside, receive and pay for all Goods on the delivery of them. But as their Demands and Expectations of Presents were very exorbitant before they would sign the Contract, or agree with us, they demanded and insisted on twenty Firelocks, and two Barrels of Powder, telling us what a mighty Service it would be to their Sultan towards reducing the Rebels, and obliging them to bring down great Store of Pepper, which they had hoarded up, and would soon enable 'em to load our Ships, and that they would pay us any reasonable Price for them. When we had agreed to this Point, and thought that all Matters had been agreed on, there arose another Difficulty, *viz.* to pay a Sooco [Suku], or quarter part of a Dollar Custom for every Pecull of Pepper (which is 132 Pounds) that we should buy. After many Debates we were forc'd to comply with this also. Then we sign'd a Contract to them in English, and they to us in the Malayo Language and Character, with the Sultan of Caytongee's Great Seal to it.

A Voyage to and from the Island of Borneo, in the East Indies, T. Warner and J. Batley, London, 1718, pp. 34–66.

6
A Visit to the Lundu Dayaks in Western Sarawak

FRANK S. MARRYAT

Frank Marryat served as a midshipman on the man-of-war HMS Samarang, *captained by Sir Edward Belcher, which departed from Portsmouth on 25 January 1843 on her last surveying cruise, sailing around the Cape of Good Hope and arriving at Singapore on 19 June. Eventually, it anchored off Kuching in the Sarawak River on 8 July 1843. The boat then went on to Hong Kong, the Philippines, eastern Indonesia, and the Malayan Peninsula, arriving back in Sarawak in August 1844.*

Marryat was a keen observer of nature and culture. He wrote about his Bornean experiences in a clear and popular way, but perhaps of greater importance were his illustrations. Probably more than any other early traveller, Marryat provided European audiences with a series of striking images of the land and its peoples—images which have continued to be used in books on Borneo to illustrate nineteenth-century Dayak life.

This extract describes a journey undertaken in 1844 to the Lundu area to the west of Kuching. The Samarang *was in the company of James Brooke and Mr Hentig, a merchant, in a native boat, as well as Captain Henry Keppel in the corvette, HMS* Dido. *Marryat provides an early general description of the Dayaks there, referring to a range of cultural traits and objects which nineteenth-century Europeans held as typical of savage life: elaborate costumes of animal skins and bird feathers, teeth-filing, betel-nut chewing, blowpipes, swords (or parang), head-hunting, 'war dances', 'communal' houses, and jungle-hunting with spear and dog.*

The Samarang *eventually arrived back in England on 31 December 1846, after having journeyed through and surveyed large areas of the Malay–Indonesian archipelago and northwards to the East Asian region, Japan and China.*

SOON after our return from the Sakarron [Skrang] the expedition to Loondoo [Lundu] was arranged, and we started in the barge and gig, accompanied by Captain Keppell in his own boat, and Mr Brooke and Hentig in one of the native boats, called a Tam-bang. The distance was about forty miles, and we should have arrived at four o'clock in the afternoon, but, owing to the narrowness of the channel, and a want of knowledge of the river, we grounded on the flats, where we lay high and dry for the space of four hours. Floating with the following tide, we discovered the proper channel, and found our way up the river, although the night was dark as pitch: when near the town, we anchored for daylight.

I may as well here give a slight description of the scenery on the Borneo rivers, all of which, that we have visited, with the exception of the Bruni, bear a close resemblance to each other. They are far from picturesque or beautiful, for the banks are generally low, and the jungle invariably extends to the water's edge. For the first fifteen or twenty miles the banks are lined with the nepa*-palms, which then gradually disappear, leaving the mangrove alone to clothe the sides of the stream. When you enter these rivers, it is rare to see any thing like a human habitation for many miles; reach after reach, the same double line of rich foliage is presented, varying only in the description of trees and bushes as the water becomes more fresh; now and then a small canoe may be seen rounding a point, or you may pass the stakes which denote that formerly there had been a fishing station. At last a hut appears on the bank, probably flanked with one or two Banana trees. You turn into the next reach and suddenly find yourself close to one or more populous and fortified towns. As you ascend higher the scenery becomes much more interesting and varied from the mangroves disappearing. Few of the rivers of Borneo are more than eighty miles in extent. The two rivers of Bruni and Coran are supposed to meet in the centre of the island, although for many miles near their source they are not much wider than a common ditch.

Before day-light of the following morning our slumbers were disturbed by the crowing of a whole army of cocks, which assured us of the proximity of the town we were in search of. We got under weigh, and, rounding the point, Loondoo hove in sight, a fine town, built in a grove of cocoa-nut trees, and by no means despicably fortified. We found our progress arrested by a boom composed of huge trees fastened together by coir cables, and extending the whole width of the river. Had our intentions been hostile, it would have taken some time to have cut the fastenings of this boom, and we should, during the operation, have been exposed to a double line of fire from two forts raised on each side of the river. The Chief of Loondoo had, however, been duly advised of our intended visit, and as soon as our boats were seen from the town, a head-man was sent out in a canoe to usher us in. After a little delay we got the barge within the boom. When within, we found that we had further reason to congratulate ourselves that we came as friends, as the raking fire from the forts would have been most effectual,

*Nipah.

for we discovered that we had to pass an inner boom equally well secured as the first. The town was surrounded by a strong stockade made of the trunks of the knee-bone* palm, a wood superior in durability to any known. This stockade had but one opening of any dimensions. A few strokes of the oars brought us abreast of it, and we let go our anchors. The eldest son of the Chief came to us immediately in a canoe. He was a splendidly formed young man, about twenty-five years old. He wore his hair long and flowing, his countenance was open and ingenuous, his eyes black and knowing. His dress was a light blue velvet jacket without sleeves, and a many-coloured sash wound round his waist. His arms and legs, which were symmetrical to admiration, were naked, but encircled with a profusion of heavy brass rings. He brought a present of fowls, cocoa-nuts, and bananas to Mr Brooke from his father, and an invitation for us to pay him a visit at his house whenever we should feel inclined.

Preparatory to landing, we began performing our ablutions in the boat, much to the amusement and delight of the naked groups of Dyaks who were assembled at the landing place, and who eyed us in mute astonishment. The application of a hair brush was the signal for a general burst of laughter, but cleaning the teeth with a tooth brush caused a scream of wonder, a perfect yell, I presume at our barbarous customs. There were many women among the groups; they appeared to be well made, and more than tolerably good looking. I need not enter into a very minute description of their attire, for, truth to say, they had advanced very little beyond the costume of our common mother Eve. We were soon in closer contact with them, for one of our party throwing out of the boat a common black bottle, half a dozen of the women plunged into the stream to gain possession of it. They swam to the side of our boat without any reserve, and then a struggle ensued as to who should be the fortunate owner of the prize. It was gained by a fine young girl of about seventeen years of age, and who had a splendid pair of black eyes. She swam like a frog, and with her long hair streaming in the water behind her, came pretty well up to our ideas of a mermaid.

As we had contrived to empty a considerable number of

*Nibong.

these bottles during our expedition, they were now thrown overboard in every direction. This occasioned a great increase of the floating party, it being joined by all the other women on the beach, and for more than half an hour we amused ourselves with the exertions and contentions of our charming naiads, to obtain what they appeared to prize so much; at last all our empty bottles were gone, and the women swam on shore with them, as much delighted with their spoil as we had been amused with their eagerness and activity.

About 10 o'clock we landed, and proceeded to pay our visit to the Chief. We were ushered into a spacious house, built of wood and thatched with leaves, capable of containing at least 400 people. The Chief was sitting on a mat with his three sons by his side, and attended by all his warriors. The remainder of the space within was occupied by as many of the natives as could find room; those who could not, remained in the court-yard outside. The Chief, who was a fine looking grey-bearded man of about sixty years of age, was dressed in velvet, and wore on his head a turban of embroidered silk. The three sons were dressed in the way I have already described the one to have been who came to us in the canoe. Without exception, those three young men were the most symmetrical in form I have ever seen. The unrestrained state of nature in which these Dyaks live, gives to them a natural grace and an easiness of posture, which is their chief characteristic. After the usual greetings and salutations had been passed through, we all sat down on mats and cushions which had been arranged for us; a short conversation with Mr Brooke, who speaks the language fluently, then took place between him and the Chief, after which refreshments were set before us. These consisted of various eatables and sweetmeats made of rice, honey, sugar, flour, and oil; and although very simple as a confectionery, they were very palatable. We remained with the Chief about an hour, and before we went away he requested our company in the evening, promising to treat us with a Dyak war dance. We took our leave for the present, and amused ourselves with strolling about the town. I will take this opportunity of making known some information I have at this and at different times obtained relative to this people.

The villages of the Dyaks are always built high up, near the source of the rivers, or, should the river below be occupied by the piratical tribes, on the hills adjoining to the source. Their houses are very large, capable of containing two hundred people, and are built of palm leaves. A village or town may consist of fifteen or twenty houses. Several families reside in one house, divided from each other by only a slight partition of mats. Here they take their meals, and employ themselves, without interfering with each other. Their furniture and property are very simple, consisting of a few cooking utensils, the paddles of their canoes, their arms, and a few mats.

In all the Dyak villages every precaution is taken to guard against surprise. I have already described the strength and fortifications of Loondoo, and a similar principle is every where adopted. The town being built on the banks of the river, the boom I have described is invariably laid across the stream to prevent the ascent of boats. Commanding the barriers, one or more forts are built on an eminence, mounting within them five or six of the native guns, called leilas. The forts are surrounded by a strong stockade, which is surmounted by a cheveaux-de-frise of split bamboos. These stockaded forts are, with the houses and cocoa nuts adjoining, again surrounded by a strong stockade, which effectually secures them from any night attack.

Great respect is paid to the laws and to the mandates of their Chiefs, although it but too often happens that, stimulated by revenge, or other passions, they take the law into their own hands; but if crimes are committed, they are not committed without punishment following them, and some of their punishments are very barbarous and cruel: I have seen a woman with both her hands half-severed at the wrists, and a man with both his ears cut off.

The religious ideas of the Dyaks resemble those of the North American Indians: they acknowledge a Supreme Being, or 'Great Spirit'; they have also some conception of an hereafter. Many of the tribes imagine that the great mountain Keney Balloo [Kinabalu] is a place of punishment for guilty departed souls. They are very scrupulous regarding their cemeteries, paying the greatest respect to the graves of their ancestors. When a tribe quits one place to reside at another, they exhume the bones of their relations, and take them with them.

'Exterior of a Sea Dyak Long-house', from William T. Hornaday, *Two Years in the Jungle: The Experiences of a Hunter and Naturalist in India, Ceylon, The Malay Peninsula and Borneo*, Kegan Paul, Trench and Co., London, 1885.

I could not discover if they had any marriage ceremony, but they are very jealous of their wives, and visit with great severity any indiscretion on their parts.

The Dyaks live principally upon rice, fish, and fruit, and they are very moderate in their living. They extract shamshoo from the palm, but seldom drink it. Their principal luxury consists in the chewing the betel-nut and chunam; a habit in which, like all the other inhabitants of these regions, from Arracan down to the island of New Guinea, they indulge to excess. This habit is any thing but becoming, as it renders the teeth quite black, and the lips of a high vermilion, neither of which alterations is any improvement to a copper-coloured face.

They both chew and smoke tobacco, but they do not use pipes for smoking; they roll up the tobacco in a strip of dried leaf, take three or four whiffs, emitting the smoke through their nostrils, and then they extinguish it. They are fond of placing a small roll of tobacco between the upper lip and gums, and allow it to remain there for hours. Opium is never used by them, and I doubt if they are acquainted with its properties.

They seldom cultivate more land than is requisite for the rice, yams, and sago for their own consumption, their time being chiefly employed in hunting and fishing. They appear to me to be far from an industrious race of people, and I have often observed hundreds of fine-looking fellows lolling and sauntering about, seeming to have no cares beyond the present. Some tribes that I visited preferred obtaining their rice in exchange from others, to the labour of planting it themselves. They are, in fact, not agriculturally inclined, but always ready for barter.

They are middle-sized, averaging five feet five inches, but very strong-built and well-conditioned, and with limbs beautifully proportioned. In features they differ very much from the piratical inhabitants of these rivers. The head is finely formed, the hair, slightly shaven in front, is all thrown to the back of the head; their cheekbones are high, eyes small, black and piercing, nose not exactly flat—indeed in some cases I have seen it rather aquiline; the mouth is large, and lips rather thick, and there is a total absence of hair on the face and eyebrows. Now the above description is not very much unlike that of an African; and yet they are very unlike, arising, I believe, from the very pleasing and frank expression of their countenances, which is their only beauty. This description, however, must not be considered as applicable to the whole of these tribes,—those on the S. E. coast of the island being by no means so well-favoured.

The different tribes are more distinguishable by their costumes than by their manners. The Dyaks of Loondoo are quite naked, and cover the arms and legs with brass rings. Those of Serebis [Saribas] and Linga are remarkable for wearing as many as ten to fifteen large rings in their ears. The Dusums [Dusuns], a tribe of Dyaks on the north coast, wear immense rings of solid tin or copper round their hips and shoulders, while the Saghai [Segai] Dyaks of the S. E. are dressed in tigers' skins and rich cloth, with splendid head-dresses, made out of monkeys' skins and the feathers of the Argus pheasant.

The invariable custom of filing the teeth sharp, combined with the use of the betel-nut turning them quite black, gives their profile a very strange appearance. Sometimes they render their teeth concave by filing.

Their arms consist of the blow-pipe (sum-pi-tan), from which they eject small arrows, poisoned with the juice of the upas; a long sharp knife, termed pa-rang; a spear, and a shield. They are seldom without their arms, for the spear is used in hunting, the knife for cutting leaves, and the sumpi-tan for shooting small birds. Their warfare is carried on more by treachery and stratagem than open fighting—they are all warriors, and seldom at peace. The powerful tribes which reside on the banks of the river generally possess several war prahus, capable of holding from twenty to thirty men, and mounting a brass gun (leila) on her bows, carrying a ball of one to two pounds weight. These prahus, when an expedition is to be made against a neighbouring tribe, are manned by the warriors, one or two of the most consequential men being stationed in each prahu. Before they start upon an expedition, like the North American Indians, they perform their war dance.

'Dyaks in Their War Dress', *Illustrated London News*, 5 November 1864.

Should their enemies have gained intelligence of the medi-tated attack, they take the precaution of sending away their women, children, and furniture, into the jungle, and place men in ambush on the banks of the river, who attack the assailants as they advance. The Dyaks are all very brave, and fight desperately, yelling during the combat like the American Indians. The great object in their combats is to obtain as many of the heads of the party opposed as possible; and if they succeed in their surprise of the town or village, the heads of the women and children are equally carried off as trophies. But there is great difficulty in obtaining a head, for the moment that a man falls every effort is made by his own party to carry off the body, and prevent the enemy from obtaining such a trophy. If the attacking party are completely victorious, they finish their work of destruction by setting fire to all the houses, and cutting down all the cocoa-nut trees; after which they return home in triumph with their spoil. As soon as they arrive another war dance is performed; and after making very merry, they deposit the heads which they have obtained in the head-house. Now, putting scalps for heads, the reader will perceive that their customs are nearly those of the American Indians.

Every Dyak [Bidayuh] village has its head-house: it is gener-ally the hall of audience as well. The interior is decorated with heads piled up in pyramids to the roof: of course the greater the number of heads the more celebrated they are as warriors.

The women of the north-east coast are by no means bad-looking, but very inferior to the mountain Dyaks before described. I have seen one or two faces which might be con-sidered as pretty. With the exception of a cloth, which is secured above the hips with a hoop of rattan, and descends down to the knees, they expose every other portion of their bodies. Their hair, which is fine and black, generally falls down behind. Their feet are bare. Like the American squaws, they do all the drudgery, carry the water, and paddle the canoes. They generally fled at our approach, if we came unexpectedly. The best looking I ever saw was one we cap-tured on the river Sakarron. She was in a dreadful fright, expecting every moment to be killed, probably taking it for granted that we had our head-houses to decorate as well as their husbands. While lying off the town of Baloongan [Balungan], expecting hostilities to ensue, we observed that the women who came down to fill their bamboos with water were all armed.

And now to resume the narrative of our proceedings:

I stated that after our interview with the old chief, and promising to return in the evening to witness a war dance, we proceeded on a stroll, accompanied by the chief's eldest son, who acted as our guide, and followed by a large party of the natives. We first examined the forts: these were in a tolerable state of efficiency, but their gun-powder was coarse and bad. We next went over the naval arsenal, for being then at peace with every body, their prahus were hauled up under cover of sheds. One of them was a fine boat, about forty feet long, mounting a gun, and capable of containing forty or fifty men. She was very gaily decorated with paint and feathers, and had done good service on the Sakarron river in a late war. These war prahus have a flat strong roof, from which they fight, although they are wholly exposed to the spears and arrows of the enemy.

We then invaded their domestic privacy, by entering the houses, and proceeded to an inspection of the blacksmith's shop, where we found the chief's youngest son, with his velvet jacket thrown aside, working away at a piece of iron, which he was fashioning into a pa-rang, or Dyak knife. The Dyak pa-rang has been confounded with the Malay kris, but they differ materially. The Dyaks, I believe, seldom use the kris, and the Malays never use the knife; and I observed, when we visited the south coast of Borneo, that the knife and other arms of the tribes inhabiting this portion, were precisely similar to those of the Dyaks on the northern coast. Customs so universal and so strictly adhered to proves not only individuality, but antiquity. Having examined every thing and every body, we were pretty well tired, and were not sorry that the hour had now arrived at which we were again to repair to the house of the rajah.

On our arrival we found the rajah where we left him, and all the chief men and warriors assembled. Refreshments had been prepared for us, and we again swallowed various mysterious confections, which, as I before observed, would have been very good if we had been hungry. As soon as the eatables had been despatched, we lighted our cheroots, and having, by a dexterous and unperceived application out of a brandy bottle, succeeded in changing the rajah's lemonade into excellent punch, we smoked and drank until the rajah requested to know if we were ready to witness the promised war dance. Having expressed our wishes in the affirmative, the music struck up; it consisted of gongs and tom-toms. The Malay gong, which the Dyaks also make use of, is like the Javanese, thick with a broad rim, and very different from the gong of the Chinese. Instead of the clanging noise of the latter, it gives out a muffled sound of a deep tone. The gong and tom-tom are used by the Dyaks and Malays in war, and for signals at night, and the Dyaks procure them from the Malays. I said that the music struck up, for, rude as the instruments were, they modulate the sound, and keep time so admirably, that it was any thing but inharmonious.

A space was now cleared in the centre of the house, and two of the oldest warriors stepped into it. They were dressed in turbans, long loose jackets, sashes round their waists descending to their feet, and small bells were attached to their ankles. They commenced by first shaking hands with the rajah, and then with all the Europeans present, thereby giving us to understand, as was explained to us, that the dance was to be considered only as a spectacle, and not to be taken in its literal sense, as preparatory to an attack upon us, a view of the case in which we fully coincided with them.

This ceremony being over, they rushed into the centre and gave a most unearthly scream, then poising themselves on one foot they described a circle with the other, at the same time extending their arms like the wings of a bird, and then meeting their hands, clapping them and keeping time with the music. After a little while the music became louder, and suddenly our ears were pierced with the whole of the natives present joining in the hideous war cry. Then the motions and the screams of the dancers became more violent, and every thing was working up to a state of excitement by which even we were influenced. Suddenly a very unpleasant odour pervaded the room, already too warm from the numbers it contained. Involuntarily we held our noses, wondering what might be the cause, when we perceived that one of the warriors had stepped into the centre and suspended round the shoulders of each dancer a human head in a wide meshed basket of rattan. These heads had been taken in the late Sakarron business, and were therefore but a fortnight old. They were encased in a wide net work of rattan, and were ornamented with beads. Their stench was intolerable, although, as we discovered upon after examination, when they were suspended against the wall, they had been partially baked and were quite black. The teeth and hair were quite perfect, the features somewhat shrunk, and they were altogether very fair specimens of pickled heads; but our worthy friends required a lesson from the New Zealanders in the art of preserving. The appearance of the heads was the signal for the music to play louder, for the war cry of the natives to be more energetic, and for the screams of the dancers to be more piercing. Their motions now became more rapid, and the excitement in proportion. Their eyes glistened with unwonted brightness. The perspiration dropped down their faces, and thus did yelling, dancing, gongs, and tom-toms become more rapid and more violent every minute, till the dancing warriors were ready to drop. A farewell yell, with emphasis, was given by the surrounding warriors; immediately the music ceased, the dancers disappeared, and the tumultuous excitement and noise was succeeded by a dead silence. Such was the excitement communicated, that when it was all over we ourselves remained for some time panting to recover our breath. Again we lighted our cheroots and smoked for a while the pipe of peace.

A quarter of an hour elapsed and the preparations were made for another martial dance. This was performed by two of the rajah's sons, the same young men I have previously made mention of. They came forward each having on his arm one of the large Dyak shields, and in the centre of the cleared space were two long swords lying on the floor. The ceremony of shaking hands, as described preparatory to the former dance, was first gone through; the music then struck up and they entered the arena. At first they confined themselves to evolutions of defence, springing from one side to the other with wonderful quickness, keeping their shields in front of them, falling on one knee and performing various feats of agility. After a short time, they each seized a sword, and then the display was very remarkable, and proved what ugly customers they must be in single conflict. Blows in every direction, feints of every description, were made by both, but invariably received upon the shields. Cumbrous as these shields were, no opening was ever left, retreating, pursuing, dodging, and striking, the body was never exposed.

Occasionally, during this performance, the war cry was given by the surrounding warriors, but the combatants held their peace; in fact they could not afford to open their mouths, lest an opening should be made. It was a most masterly perform-ance, and we were delighted with it.

As the evening advanced into night, we had a sort of extemporary drama, reminding us of one of the dances, as they are called, of the American Indians, in which the war-riors tell their deeds of prowess. This was performed by two of the principal and oldest warriors, who appeared in long white robes, with long staves in their hands. They paraded up and down the centre, alternately haranguing each other; the subject was the praise of their own rulers, a relation of their own exploits, and an exhortation to the young warriors to emulate their deeds. This performance was most tedious; it lasted for about three hours, and, as we could not understand a word that was said, it was not peculiarly interesting. It, however, had one good effect: it sent us all asleep. I fell asleep before the others, I am told; very possible. I certainly woke up the first, and on waking, found that all the lights were out, and that the rajah and the whole company had dis-appeared, with the exception of my European friends, who were all lying around me. My cheroot was still in my mouth, so I re-lighted it and smoked it, and then again lay down by the side of my companions. Such was the wind-up of our visit to the rajah, who first excited us by his melodramas, and then sent us to sleep with his recitations.

The next morning, at daylight, we repaired to our boats, and when all was ready took leave of the old rajah. The rajah's eldest son had promised to accompany us to the mouth of the river, and show us how the natives hunted the wild pigs, which are very numerous in all the jungles of Borneo.

We got under weigh and proceeded down the river accom-panied by a large canoe, which was occupied by the rajah's son, six or seven hunters, and a pack of the dogs used in hunting the wild boar on this island. These dogs were small, but very wiry, with muzzles like foxes, and curling tails. Their hair was short, and of a tan colour. Small as they are, they are very bold, and one of them will keep a wild pig at bay till the hunters come up to him.

We arrived at the hunting ground at the mouth of the river

in good time, before the scent was off, and landed in the Tam-bang. Our captain having a survey to make of an island at the mouth of the river, to our great delight took away the barge and gig, leaving Mr Brooke, Hentig, Captain Keppell, Adams, and myself, to accompany the rajah's son. Having arranged that the native boat should pull along the coast in the direction that we were to walk, and having put on board the little that we had collected for our dinners, we shouldered our guns and followed the hunters and dogs. The natives who accompanied us were naked, and armed only with a spear. They entered the jungle with the dogs, rather too fatiguing an exercise for us, and we contented ourselves with walking along the beach abreast of them, waiting very patiently for the game to be started. In a very few minutes the dogs gave tongue, and as the noise continued we presumed that a boar was on foot; nor were we wrong in our conjecture; the barking of the dogs ceased, and one of the hunters came out of the jungle to us with a fine pig on his back, which he had transfixed with his spear. Nor were we long without our share of the sport, for we suddenly came upon a whole herd which had been driven out of the jungle, and our bullets did execution. We afterwards had more shots, and with what we killed on the beach, and the natives secured in the jungle, as the evening advanced we found ourselves in possession of eight fine grown animals. These the rajah's son and his hunters very politely requested our acceptance of. We now had quite sufficient materials for our dinner, and as we were literally as hungry as hunters, we were most anxious to fall to, and looked upon our pigs with very cannibal eyes. The first thing necessary was to light a fire, and for the first time I had an opportunity of seeing the Dyak way of obtaining it. It differs slightly from the usual manner, and is best explained by a sketch. Captain Keppell, who was always the life and soul of every thing, whether it was a fight or a pic nic, was unanimously elected caterer, and in that capacity he was most brilliant. I must digress a little to bestow upon that officer the meed of universal opinion; for his kindness, mirth, and goodness of heart, have rendered him a favourite wherever he has been known, not only a favourite with the officers, but even more so, if possible, with the men. In the expeditions in which Keppell has been commanding officer, where the men were worn out with continued exertion at the oar, and with the many obstacles to be overcome, Keppell's voice would be heard, and when heard, the men were encouraged and renewed their endeavours. Keppell's stock, when provisions were running short, and with small hopes of a fresh supply, was freely shared among those about him, while our gallant captain, with a boat half filled with his own hampers, would see, and appeared pleased to see, those in his company longing for a mouthful which never would be offered. If any of the youngsters belonging to other ships were, from carelessness or ignorance, in trouble with the commanding officers, it was to Keppell that they applied, and it was Keppell who was the intercessor. In fact, every occasion in which kindness, generosity, or consideration for others could be shown, such an opportunity was never lost by Keppell, who, to sum up, was a beloved friend, a delightful companion, and a respected commander. As soon as our fire was lighted, we set to, under Keppell's directions, and, as may be supposed, as we had little or nothing else, pork was our principal dish. In fact, we had pig at the top, pig at the bottom, pig at the centre, and pig at the sides. A Jew would have made but a sorry repast, but we, emancipated Christians, made a most ravenous one, defying Moses and all his Deuteronomy. We had plenty of wine and segars, and soon found ourselves very comfortably seated on the sand, still warm from the rays of the burning mid-day sun. Towards the end of a long repast we felt a little chilly, and we therefore rose and indulged in the games of leap-frog, fly-the-garter, and other venturous amusements. We certainly had in our party one or two who were as well fitted to grace the senate as to play at leap-frog, but I have always observed that the cleverest men are the most like children when an opportunity is offered for relaxation. I don't know what the natives thought of the European Rajah Brooke playing at leap-frog, but it is certain that the rajah did not care what they thought. I have said little of Mr Brooke, but I will now say that a more mild, amiable, and celebrated person I never knew. Every one loved him, and he deserved it.

After we had warmed ourselves with play, we lighted an enormous fire to keep off the mosquitoes, and made a bowl of grog to keep off the effects of the night air, which is occasionally very pernicious. We smoked and quaffed, and had many a merry song and many a witty remark, and many a laugh about nothing on that night. As it is highly imprudent to sleep in the open air in Borneo, at ten o'clock we broke up and went to repose in the boats under the spread awnings. Just as we were selecting the softest plank we could find for a bed, we had an alarm which might have been attended with fatal consequences. I omitted to mention that when we rose to part and go into the boats, one of the party threw a lighted brand out of the fire at the legs of another; this compliment was returned, and as it was thought very amusing, the object being to leap up and let the brand pass between your legs, by degrees all the party were engaged in it, even the rajah and the natives joined in the sport, and were highly amused with it, although with bare legs they stood a worse chance of being hit than we did. At last the brands were all expended and the fire extinct, and then, as I said, we went away to sleep under the boats' awnings. We were in the act of depositing our loaded rifles by our sides in a place of security, when the unearthly war cry rose in the jungle, and in the stillness of the night these discordant screams sounded like the yelling of a legion of devils. Immediately afterwards a body of natives rushed from the jungle in the direction of the boats, in which we supposed that our European party were all assembled. Always on our guard against treachery, and not knowing but that these people might belong to a hostile band, in an instant our rifles were in our hands and pointed at the naked body of natives, who were now within twenty yards of us. Mr Hentig was on the point of firing, when loud shouts of laughter from the Dyaks arrested his hand, and we then perceived that Mr Brooke and others were with the natives, who enjoyed the attempt to intimidate us. It was fortunate that it ended as it did; for had Mr Hentig been more hasty, blood must have been shed in consequence of this native practical joke. We joined the laugh, however, laid down our rifles, then laid ourselves down, and went fast asleep, having no further disturbance than the still small voice of the mosquito, which, like that of conscience, is one that 'murders sleep'.

The following morning we bade adieu to our friendly hunting party, and I must not here omit to mention a trait of honesty on the part of the Dyaks. I had dropped my pocket handkerchief in the walk of the day before, and in the evening it was brought to me by one of the natives, who had followed a considerable distance to bring it to me. It must be known, that a coloured silk handkerchief is to one of these poor Dyaks, who are very fond of finery, an article of considerable value. He might have retained it without any fear; and his bringing it to me was not certainly with any hope of reward, as I could have given him nothing which he would have prized so much as the handkerchief itself. He was made a present of it for his honesty.

We bade farewell to our friends at Kuchin, and continued our survey on the coast. The boats were now continually employed away from the ship, which moved slowly to the westward. At this time exposure and hard work brought the fever into the ship. The barge returned in consequence of four of her men being taken with it, and our sick list increased daily. A few days afterwards the coxswain of the barge died, and was buried along side the same morning. The death, after so short an illness, damped the spirits of the officers and men, particularly of those who were ill. After this burial we sailed for Sincapore. At this time our sick report contained the names of more than thirty men, with every probability of the number being increased; but, thanks to God, from change of air, fresh provisions, and a little relaxation from the constant fatigue, the majority were in a short time convalescent. On the 25th of September we arrived at Sincapore.

Borneo and the Indian Archipelago with Drawings of Costume and Scenery, Longman, Brown, Green, and Longmans, London, 1848, pp. 71–93.

7
An Expedition against the Warlike Sea Dayaks (Ibans) of the Saribas

CHARLES BROOKE

This book is based on Charles Brooke's journals, written from 1852 to 1863. He was born in 1829, nephew of the first Rajah of Sarawak, Sir James Brooke. At the age of twelve Charles joined the Royal Navy. In 1844 he served as midshipman of HMS Dido, *captained by Henry Keppel; the* Dido *sailed in Sarawak waters in the 1840s. In 1852, as a very young man, Charles entered the service of Sir James Brooke in Sarawak.*

In the late 1850s and into the 1860s Charles Brooke was in the frontline of the Raj's pacification campaigns against the warlike Ibans of the Skrang and Saribas Rivers. Although he was a resident in Sarawak, serving in its administration, he travelled widely in the region. The following passage has been deliberately chosen because it illustrates a rather different aspect of travel, and that is a journey into battle. This is not travel for pleasure, curiosity, or scientific discovery, but for physical confrontation. After 1853, when the British Royal Navy withdrew its military support from James's regime, the fragile personal rule of the Brookes was left to its own devices. The Rajah's nephew was a tower of strength in this situation, and as Reece notes, Charles Brooke performed the role of 'a Dayak warrior chief' (pp. vi, viii).[1] He led punitive forces of Malays and friendly Dayak mercenaries against other 'rebellious' warring tribes.

Charles Brooke, who succeeded his uncle Sir James in 1868, as Second Rajah of Sarawak, reigned for forty-nine years until his death in 1917. He very much stamped his personality and authority on Sarawak during those years, but it must be as a leader of native armed forces and as a soldier in action when we see Charles Brooke at his best.

The following extract describes an expedition of pacification against the Ibans or Sea Dayaks of the Saribas in April 1858. Charles Brooke was very close to his Ibans; he loved their culture, personality, and their company, and he believed in the use of armed force only as a last resort. But as we see, he was quite prepared, if necessary, to extend Brooke rule against the Ibans by military means.

APRIL, 1858.—I had for many months been tormented by the affairs in Saribus [Saribas], which had been for generations the hot-bed of head-hunting and piracy in every shape. The people were now—partly owing to the Chinese insurrection—becoming much more audacious, and

---

[1]R. H. W. Reece, 'Introduction' to Charles Brooke, *Ten Years in Sarawak*, Oxford University Press reprint, Singapore, 1990, pp. v-xxv.

I found it had been to no purpose holding communication with even the Malays, who a few days ago refused to receive a letter, and declared they intended shortly to ascend the river, and live with the Dyaks, and eat pork as they did. So it was evident a crisis was approaching which would require resolute action, or our prestige would be injured in this quarter. This we could by no means afford to lose, as stoppage of all trade and communication on the coast would inevitably ensue. Two Dyak friends (the likeness of one is in Mr Spenser St. John's book), named Bakir and Ejau Umbul, came over from that river, and brought intelligence that a fleet of forty large Dyak boats were now ready on the Saribus side, commanded by Lintong, and were only waiting for Sadji, whose boat was building at Paku, about forty miles from the mouth, but was not yet completed. Their intention was to go to sea, and take heads in any direction. After thinking over what was best to be done, I resolved to send an express-boat to Sarawak for reinforcements, and at once set to work to put my dilapidated flotilla into order. I was sadly in want of large boats, having only one fit for use. The Dyaks, I was well aware, would be ready directly, but their force could not be depended upon, except in an irregular fashion.

In three days the big boat was ready, with a 3-pounder in her bows, and sixty men aboard, well armed and provisioned, and we set off. On reaching Lingga, luck favoured us considerably, for here we found our small gunboat schooner, 'Jolly Bachelor', commanded by John Channon, had just arrived, having brought some passengers for Banting. I now felt considerably relieved in my mind at having an extra force, which I detained for this important service.

After embarking, I started with a picked crew, and the numerous boats were following in our track. We entered Saribus unknown to all the population, the greater part of which were residing near the mouth. A boat, containing some of the head men, came alongside of us as we were passing up. Agitation was strongly depicted on their countenances, and they whispered inquiries of some of the crew whether they were to be attacked as well as the Dyaks. I informed them that they might follow if they were inclined, but that I should ask nothing of them. A short distance higher up we anchored off another village, the chief of which was favourably disposed to us. He at once came on board, when I

'Surprise of the Pirate Village of Kenowit', from Rodney Mundy, *Narrative of Events in Borneo and Celebes, down to the Occupation of Labuan: From the Journals of James Brooke Esq.*, John Murray, London, 1848.

told him I had perfect confidence in him, and wished him to take charge of the vessel, to pilot her up past the dangerous shoals of the river. He looked anxious, but was evidently proud of being the individual singled out for this duty, and promised to do his best. He pointed to his son, a fine strapping youth of seventeen years, and begged me to make every use of him. Our party of boats assembled, numbering only one hundred as yet. At an early hour the next morning we weighed. The river was broad and deep, and the tide whirled us up at a sweeping pace. When we brought up below the point where the bore first rises, there was some danger of being ripped open by the cable as we swerved about in an eddy which swung us round and round, and heeled the vessel over considerably; then a back-water would take the bow, bringing it up with a jerk against the cable. We started again next morning, and soon came to the dangerous shoals; one touch on them would have rolled us over like nine-pins. The tide was making about eight miles an hour, and we had a few boats towing, to give us steerage-way. At one time there was only one foot of water to spare under the keel, and then Channon's face looked sadly anxious; but the lead was useless at such a time, and the pilot must trust solely to his knowledge of the different sands and difficulties.

News was brought us that Sadji had abandoned his boat, and had retreated up the river, and the inhabitants higher up were preparing their wives and families for a retreat into the interior, leaving only the men to defend the houses and property. They had expected to find me in a small boat as usual, when they would have made an attack with their heavy force, but now, they said, I had come in a house, for so they styled our small pinnace.

We passed most dangerous places without an accident, and then anchored again for the night: the vessel ranged about to such a degree as to keep us awake the greater part of it. To my great disappointment I found the cholera had followed us, and three boats' crews had already come alongside asking for medicine; two poor fellows in one boat had the complaint in the acutest form, and were suffering most excruciating pain from cramp. I administered an almost neverfailing remedy, 'The Bishop of Labuan's Pill', and rubbed the men with Kaya putih oil; they were better before morning, and so were all those who could take these remedies in time; but, alas, many did not, and died ere the morning sun arose. We lifted anchor at sunrise, and a number of friendly Dyaks came on board, or followed close to us, dressed in their finest clothes.

All real dangers in the river were now passed. Hilly banks were on either side, and every here and there was a Dyak house, out of which women and children peeped, and pretty laughing girls in brightly coloured costumes waved their hands, and seemed rejoiced at seeing us, asking us to stop in their most coquettish manner. With a strong tide we swept past beautiful farming grounds heavily laden with padi, waving yellow and ripe in the morning sun; and at 9 A.M. anchored at the mouth of the Padeh river, which is narrow and shallow. There was no room for swinging, so we moored head and stern, and hauled our anchor up, and hung it on to the branch of a high fruit tree, to keep the chain from fouling with drift-wood. Luxuriant fruit trees were grouped around us on either side, that of the luscious durian being most plentiful, and very magnificent in size. There was great commotion among the inhabitants, who evidently had not made up their minds what to do, whether to flee or stay; all our old-established enemies were far away, but there were many others who hovered half way. Some were trying to pass with goods and chattels in their boats; these I seized until matters were a little more settled, and shortly our vessel's store-room was full of valuable jars. A remark passed by a Malay somewhat surprised the owners; he told them that heretofore the Saribus had had a heaven all to themselves, different to all others, and now they would find out their mistake.

We housed in and collected arms on deck, and arranged a 6-pounder gun so as to point up the river, in case any of the enemy's forty boats made their appearance; the force was collecting for the whole day, and in the evening we held a consultation to decide future proceedings. It was arranged that a party should be organised to make an attack on the Padeh stream against Nanang and Sadji's houses, which were about half-a-day's walk away. I refused to go inland unless the expedition remained away for three days, and in the morning, as everything was well prepared, and I was ready equipped, a large number of the principal men came to beg me to remain, aboard or not,—at all events, to stay a night away; so I gave up the idea of going at all, but gave directions that a party of Malays and Dyaks should go and lay the country waste within a day's march of where we then lay. In an hour they started, and we could see the line straggling far away in the distance before we lost sight of them in the jungles. The Ballau Dyaks were leading, as these were their old enemies, from whom they had suffered so much, and now was the time to avenge themselves. At mid-day they descried the roofs of the enemy's houses, and shortly afterwards the leaders were attacked by a volley of spears hurled from a hillock. Janting, with a son-in-law on each hand, advanced, followed by his people, and opposed the party with drawn swords; one of his sons cut down his man, decapitated him, and Janting himself had come in contact with another, when his other son-in-law fell with two spear wounds, and would have lost his head, if his father had not most opportunely dealt a terrific blow at his adversary, and then stood guard over his wounded relation, while the enemy had time to make off, fighting indiscriminately with our people. In this scrimmage many were wounded; the Saribus retreated, and knowing the country, were soon out of reach. There were no muskets on either side, and the Malays were far in the rear, and throughout the day evinced great fear, and did little service. An advance was then made upon the houses, which were found comparatively empty. They were gutted of the few things left in them, and burnt. Our Dyaks strayed about, and on reaching another house they rushed in, and on some of them seizing a valuable jar, the rightful owner turned round and inquired 'What they were about?' whereupon our party was attacked, and several of them killed for making this foolish mistake of thinking the inhabitants a part of their own force. However, the enemies were afterwards driven out, and their houses burnt, and our party returned victorious.

The Malays showed the most wholesome dread of the enemies, and on the previous day I had overheard one of the head men making inquiry, whether it would be possible to overcome such people as Saribus Dyaks, who had been victorious for as many generations? These Malays also were exceedingly divided as to whom they should ultimately follow—whether Dyak or white man; and were doubtful as to our sufficiently supporting and protecting them if they became really enemies of the Dyaks: however, I was now in a position to clinch their sincerity, being placed between them and the Dyak community, though if they had joined the latter they would doubtless have given us much trouble. I resolved not again to abandon this river, but to build a fort and permanently take possession in order to guard and establish some system of government, whereby inch by inch we might hold fast what we gained, and so prevent all our work from being undone by Malay rascality. In the afternoon I went ashore to choose a spot for building the fort, and cleared a large patch of ground well adapted for offensive and defensive operations, commanding the river for several miles up the reaches, and with a picturesque view on all sides.

While taking notes of this locality, a tumult arose below, which brought about silence among us; we descended to the banks of the river in all haste, scrambling through the brushwood as best we could. On reaching our force I found our Dyaks were fighting among themselves, and disputing over the head of an enemy. They were making a fearful commotion, the boats drifting across each other, and men standing with drawn swords in their hands. I saw there was little time to lose, so rushed down the mud bank to the dingy, and shoved into the midst of this promiscuous mêlée. Janting was the leader, vociferating in true Dyak fashion with the utmost exasperation. His temper was hot enough to drive him to commit any mischief when once aroused. I closed with his boat, placed my hand on his shoulder, spoke a few quiet words, asking him not to cast disgrace on the whole of the force by fighting with his own friends. He at once silently slunk inside his boat, the sounds died away, and peace was restored; but such rows are exceedingly dangerous and unpleasant. No Malay attempted to interfere, and it was only by knowing the man that I was able to succeed without resorting to severity, when one drop of blood might have led I don't know where.

I returned to the Jolly covered with mud, and after bathing in the cool stream, found Janting sitting on deck weeping like a child. He apologised for his bad behaviour, and begged to be forgiven, alleging as an excuse the excitement of the morning's encounter with the enemy, and his son-in-law's wounds; people wished, he said, to seize his property, and he defended it. I dressed the wounded man, who was in no danger; he had one job from a spear in his groin, and another through his arm, and was in much pain, but the healing powers of a native's constitution are surprising. The next morning I found our force was breaking up to return home; the cholera had attacked them very severely during the night, and several had died within a few hours. One poor fellow, whom I had known for years, called at midnight for medicine, and finding me asleep, the watchman refused to arouse me. He was dead before daylight. The Dyaks expressed their regret at being obliged to leave, but go they must. We held a consultation among the Malays about building the fort, and they unanimously recommended it to be erected further down, as here, they said, it would be next to impossible to find supplies. After weighing the pros and cons I yielded, and followed their advice, so we dropped down a few reaches into more open water. The Dyak enemies were hovering about astern of us, and some yelled and flourished their swords. The Saribus Malays proceeded to cut wood, which they were to bring for the building of the new fort. The cholera was playing sad havoc among them; they complained but little, though their faces told their misery. Fortunately the Jolly's crew had passed through their ordeal in Sarawak, and were not further molested.

The reinforcements now arrived from Sarawak in charge of Fitze Cruickshank. They were very late in coming, and no further active operations were in contemplation at present. We beached the pinnace on the bank, and monotonously awaited the wood. Old Abang Boyong had been the active working man, and was inclined to favour the Dyak cause, offering peace and reconciliation. His wish was that they should be persuaded to pay a small fine, and be received on friendly terms; but my opinion was, that they were far too strong to wish to submit. I, however, permitted Abang Boyong to do his utmost with overtures, and he had appointed an interview with the enemy at the mouth of the Padeh. On the evening of the day that the meeting was to have been held, Boyong came to me with an anxious face, and in a few words told me he had been to the spot appointed, and never had such a narrow escape of his life before. As he was speaking with two or three fellows on the ground, Sadji's party assembled in numbers, and armed, and hid themselves in the brushwood. A few minutes more, and he would have been taken captive or killed, but fortunately spying Sadji's red jacket in the distance, he had time to make off in his boat beyond their reach. He confessed he had been shamefully deceived by their artful designs, and his subdued tone had a quiet anger about it. He begged me to follow them up, if possible, or at any rate make some demonstration to disperse them.

Before the assembled chiefs in the evening we decided the way of procedure, and all were favourable except a patriarchal Hadji from Sarawak, who emphatically exclaimed, 'Remember, Tuan, whatever else occurs, do not let us have any fighting!' At an early hour on the following morning we set off in boats only, and, with a strong flood-tide, floated up the Padeh stream without making a sound. The bowsman and steersman only kept the boat in her proper position; the remainder of the crew, with their muskets, were all in readiness. Boyong was in front, and the stream was so small as only to permit one boat at a time to pass on. At last, on rounding a point, our enemies were seen sitting near the bank, and the leaders fired a volley, which was followed by a very indiscriminate blazing of musketry. Our boats became crowded together, and great confusion and noise ensued. I felt in a perilous position, and thought that every moment some of our people would shoot each other. As for an enemy, none but the leading boats saw one. Some of them rushed ashore, but soon returned. None of our party had been wounded, and our enemies were dispersed. We immediately retired, or we should have been unable to move at low water. We dropped back stern first, in the same quiet way as we had advanced.

This short trip plainly proved to me that great risk is attached to a Malay force, unless accompanied by Dyaks; for the latter act as a look-out, and are able to cope with Dyaks in jungles, where the Malays are next to useless unless they happen to have been born and bred among Dyaks.

The delay and the sickness caused the time to pass in the most irksome manner, and my condition was weak and wretched, and was made worse by living in this small cabin of the gunboat, in which one can only sit upright. I seldom passed a night without an alarm of some kind or other, and my nervous system was shaky from confinement. One night I jumped up in a dream, and pointed my rifle at a friendly boat, but awoke just in time to see my error; so, for a change, I left Fitze and John Channon in charge, and set off myself overland for Sakarang [Skrang]. It was a most fatiguing walk, without any regular path, and the greater part of the way over very steep hills, or through miry swamps, the latter knee-deep. The distance was about twelve miles, with a crooked road, and while progressing I thought that no persuasion or pecuniary reward could induce me to attempt such an experiment again. It did not cease raining the whole way, and on reaching Sakarang I was glad to find the cellar not quite empty; a bottle of sherry somewhat restored the inner and outer man.

I made the most of my one day's holiday, but had engaged to return the following day, so we trudged back again, and this time I did not feel the fatigue. I was much the better for this spurt, and had restored my nerves by the rough walking.

We now went to work at the defence, consisting of a single-roofed house planked all round, having a small aperture for an entrance, with a long notched pole suspended to answer the purpose of a ladder. This was pulled up at night, and, with ports closed, there was no opening left to admit an enemy. Four three-pounder guns were mounted in the upper story. It was soon completed, with a large space around cleared of jungle and brushwood, to prevent dangers arising from an enemy prowling about with fire or sword in the vicinity of the building. The house of a good old Dyak stood near, and his party were always ready when needed. My friend and pilot, Abang Dondang, removed his people, to take up their abode close to the fort, mustering about 130 fighting Malays. I despatched the pinnace back to Sarawak, and after seeing the place sufficiently secure, placed Fitze Cruickshank in charge of the station. He was young, but strong and plucky, with an abundance of Scotch prudence and plain common sense. After giving him the necessary instructions, I left in a small boat, and when near the mouth met two Dyak boats, one of which was towing a sampan. On my reaching Lingga the next day, I received intelligence that a party of Dyaks had been near the mouth of the Lingga, had met one boat, which they stopped with the offer to sell sirih leaf. When near, the crew drew their parangs (swords), cut down one woman, whose head they obtained, and took the daughter prisoner. The father jumped overboard in time to save his life.

These were the Dyaks we passed in Saribus river, with the captive and head aboard their boat. I deplored at the time the escape of the rascals, the head of whom was our enemy, Sadji; but subsequently have had reason to be satisfied that we passed the enemy untouched. I was afterwards told that when Sadji saw our boats he sat with his drawn sword on the captive's throat, ready to cut off her head if she had spoken, or we had taken notice of them. They would then have gone on shore with both heads, and found their way overland to their haunts, and we should not have been able to follow in the strange jungles. This act of Sadji's was the most bare-faced that had been committed for several years, and it was evidently necessary for us to take every precaution, as the fight was to be a hard one.

I returned to Sakarang, where everything looked homely

and nice; as it had ever been my particular care to keep my
abode in apple-pie order, and not to lose all habits of civilisation and neatness. My hall of audience was capable of
holding six hundred men. Along it were arranged the arms
between the wooden posts, with chairs and tables at each
end. Occasionally, when lonely, I used to invite natives to see
the magic lantern, and, with a musical box and some dancers,
an evening could be passed with moderate pleasure. The
sword-dance is excessively ungraceful and uninteresting; a
stiff mode of pirouetting round and round is the general
figure, which would be perfectly useless in actual sword-
play. Such displays did not often, however, take place; and
one's life is particularly monotonous, books alone keeping
the mind from flagging and becoming unhinged. My garden
afforded me the next greatest occupation, and plates full of
flowers were daily brought in and set on the different tables
in the apartments, wafting a delicious perfume of these
strong Eastern scented jessamines (chimpakas) through them.

The natives have no idea of the evil reported to arise from
such perfumes at night, for it is their favourite practice to
cover their pillows with opening blossoms, which are left till
they are quite faded. Most of these flowers open after sunset,
and fade soon after sunrise, unless they are plucked and kept
in a cool shady room. My cattle afforded me much interest,
and they were thriving well and multiplying fast. But the
chief charm which kept me from sinking to the depths of
despair, was my interest in the Dyaks.

Ten Years in Sarawak, Tinsley Brothers, London, 1866, pp. 269–85.

8
The Ascent of Mount Kinabalu

SPENSER ST. JOHN

*Spenser St. John provides us with a description of one of the
great human adventures of Borneo: one of the first successful
conquests of Kinabalu, the highest mountain in South-East
Asia.*

Spenser St. John is one of the most important figures in the history of nineteenth-century British imperialism in island South-East Asia. Born in 1825 he became a close personal friend of Sir James Brooke, and from 1848 served in his administration in Sarawak as the Rajah's personal secretary. He was later to write a biography of the Rajah, based on Sir James's personal papers and correspondence, and St. John's own knowledge of him. St. John succeeded Sir James Brooke as British Consul-General to the native states of Borneo, and in this capacity took up residence in Brunei in 1856. He was later knighted.

In April 1858 St. John, together with Hugh Low, the colonial treasurer of the British colony of Labuan, attempted the ascent of Kinabalu in north-east Borneo. Unfortunately, Low had to abandon the final climb because of painfully swollen feet. However, St. John went on with his guides and successfully reached one of the summits. A few months later St. John and Low climbed the mountain for a second time.

St. John writes in a calm, measured way, with hardly a mention of the difficulties and dangers of tropical life and travel. He was not a naturalist, and therefore his descriptions of fauna and flora are commonly not as informative and penetrating as those of such trained and interested observers as Beccari (Passage 3) or Wallace (Passage 9). Yet, St. John certainly demonstrates a breadth of knowledge of northern Borneo, acquired during thirteen years residence there. He was passionately devoted to exploration, and the personal qualities which he displays and his undoubted expertise on Sarawak has perhaps led Tom Harrisson to place St. John's Life in the Forests of the Far East *in the same class as Wallace's* The Malay Archipelago.[1]

St. John's description of Mount Kinabalu and his adventures there helped fix this wonder of nature firmly in the European consciousness. There is probably no other feature which best typifies northern Borneo than the sacred mountain of the local Dusun (or Kadazan) people.[2]

TO ascend Kina Balu had been an ambition of mine, even before I ever saw Borneo. To have been the first to do it would have increased the excitement and the

[1]Tom Harrisson, 'Introduction' to Spenser St. John, *Life in the Forests of the Far East*, Oxford University Press reprint, Kuala Lumpur, 1974, pp. v–xvi.

[2]St. John calls them Ida'an in his account.

pleasure. However, this satisfaction was not for me. Mr Low, colonial treasurer of Labuan, had long meditated the same scheme, and in 1851 made the attempt. It was thought at the time but little likely to succeed, as the people and the country were entirely unknown; but by determined perseverance Mr Low reached what may fairly be entitled the summit, though he did not attempt to climb any of the rugged peaks, rising a few hundred feet higher than the spot where he left a bottle with an inscription in it.

In 1856, Mr Lobb, a naturalist, reached the foot of the mountain, but was not allowed by the natives to ascend it.

In 1858, Mr Low and I determined to make another attempt; and early in April I went over from Brunei to Labuan to join him. We waited till the 15th for a vessel, which we expected would bring us a supply of shoes, but as it did not arrive we started. This was the cause of most of our mishaps,—as a traveller can make no greater mistake than being careless of his feet, particularly in Borneo, where all long journeys must be performed on foot.

*   *   *

The name of this place was Batong: from it Kina Balu bore S. E., and Saduk Saduk 15° east of south; the latter appears from this view to be a peaked mountain between 5,000 and 6,000 feet high. Kina Balu of course absorbed our attention: at night, as the sun shone brightly on its peaks, it wore a very smiling appearance. The summit seemed free from all vegetation, and streams of water were dashing over the precipices.

Started next morning at a quarter to eight, and soon arrived at a place where the river divided, the Penantaran coming from an E. N. E. direction. Its bed was full of large blocks of serpentine (though after passing the mouth of this branch we met with very few specimens of that kind of rock). There is a village of the same name as the branch close to the junction. We followed the right-hand branch—direction about south— keeping close to the banks, crossing and recrossing continu- ally, seeing occasionally a few houses. We were now passing through sandstone ranges, but the country had no remark- able features. At 9.40, stopped to breakfast, having made about four miles; our followers gradually closed up. At eleven we pushed on again. Huge granite boulders are now common, and under the shelter of one mighty stone we rested for half an hour, waiting the arrival of our straggling followers.

One of the greatest advantages of travelling with an intelligent companion is the interchange of ideas, and consequently the more accurate noting of observations. As we sat beneath the shade of the huge granite boulder, surrounded on all sides by sandstone hills, we could not but speculate how it came there. Without having recourse to the glacier theory, the reason appeared to me simple. It is evident that the level of the country was very much greater in former times than at present, and that water is the great agent by which these changes have been effected.

The streams continually cut their way deeper in the soil, as we may daily observe: the increasing steepness causes innumerable landslips, and the process going on for ages, the whole level of the country is changed, and plains are formed from the detritus at the mouths of the rivers. Huge granite masses, falling originally from the lofty summit of Kina Balu, would gradually slip or roll down the ever-forming slopes which nature is never weary of creating.

In ascending some of the steeps that rise on either side of the streams near Kina Balu, we continually came across boulders of granite, which, in comparatively few years, will, through landslips, roll many hundred feet into the stream below, to commence their gradual movement from the mountain. I have continually come across evidences of the Bornean rivers having flowed at a much higher level than at present, finding layers of water-worn pebbles, a hundred feet above the present surface of the stream. In Borneo, where the rain falls so heavily, the power of water is immense. After a heavy storm, the torrents rise in confined spaces often fifty feet within a few hours, and the rush of the stream would move any but the largest rocks, and wash away most of the effects of the landslips.

Standing on a height overlooking a large extent of country, it is instructive to be able to survey at a glance the great effect caused by the rivers and all their tributaries, deep gullies marking every spot where an accession joins the parent stream. After heavy rains, the rivers present the colour of *café au lait*, from the large amount of matter held in temporary suspension, and on taking out a glassful, I have been surprised by the amount of sediment which has immediately fallen to the bottom.

The walk was becoming rather tiring; drizzling rain rendering the stones very slippery, and having continually to make the mountain torrent our path, it was severe work for our bare feet. The rain continuing, and the stream rapidly rising, we halted at some farmhouses in the midst of a long rice-field. Fording the river is difficult work; the water rushing down at headlong speed, renders it necessary to exert one's utmost strength to avoid being carried away: the pole in both hands, placed well to seaward, one foot advanced cautiously before the other, to avoid the slippery rocks and loose stones. I found that this fatigued me more than the walking. The water became much cooler as we approached the mountain, while the land is rapidly increasing in elevation. The river was full of Ida'an fish-traps, made by damming up half the stream, and forcing the water and fish to pass into a huge bamboo basket. They appeared to require much labour in the construction, particularly in the loose stone walls or dams. As we advanced, we found the whole stream turned into one of these traps, in which they captured very fine fish, particularly after heavy rain. I bought one with large scales, about eighteen inches long, which was of a delicious flavour.

To see the young Ida'an ford the stream, raised both my envy and my admiration; with the surging waters reaching to their armpits, with a half-dancing motion, they crossed as if it were no exertion at all. So much for practice. During the last three hours we did not make more than four miles, though out of the stream the paths were good. The rain continuing to pour heavily, we determined to stop, as I have said, at these Ida'an huts, which were situated opposite the landing-place of the village of Tambatuan, concealed by the brow of a steep hill rising on the other bank. We sent a party there to buy rice, which became cheaper as we advanced: these villages also possessed abundance of cattle and buffaloes. We were much pleased to find the great confidence shown by the people; we often met parties of women and girls, and on no occasion did they run away screaming at the unusual sight of a white face. Several of them came this afternoon to look at us, and remained quite near for some time, interested in watching our proceedings. Kina Balu was cloud-hidden this evening.

During the night our rest was much disturbed by bees, who stung us several times, and Mr Low, with that acuteness which never deserts him in all questions of natural history, pronounced them to be the 'tame' bees, the same as he had last seen thirteen years ago among the Senah Dayaks in Sarawak. About mid-night we were visited by a big fellow, who, our guides assured us, wanted to pilfer; but we found next morning that he had come to complain of his hives having been plundered. On inquiry, we discovered the man who had done the deed. He was fined three times the value of the damage, and the amount handed over to the owner.

A great many questions were asked as to what could be our object in visiting Kina Balu: to tell them that it was for curiosity would have been useless: to say that we were seeking new kinds of ferns, pitcher-plants, or flowers, would not have been much more satisfactory to them. Some thought we were searching for copper or for gold, while others were equally convinced we were looking for precious stones. One man sagaciously observed that we were seeking the *Lagundi* tree, whose fruit, if eaten, would restore our youth and enable us to live for countless years, and that tree was to be found on the very summit of Kina Balu. To-day an Ida'an came, I suppose to try us, and said he knew of copper not more than half a day's journey from our path, and offered to take us to it; seeing we were not to be tempted, another told us of a tree of copper that was to be found a few miles off; but even that did not alter our determination to make the best of our way to the mountain. We left the questioners sadly puzzled as to what possibly could be our object in ascending Kina Balu.

All the Bajus [Bajaus] and Borneans are convinced that there is a lake on the very summit of this mountain, and ask, if it be not so, how is it that continual streams of water flow down its sides. They forget that very few nights pass without there being rain among the lofty crags, even when it is dry on the plains. Sometimes the sun, shining on particular portions of the granite, gives it an appearance of great brilliancy; and those who formerly ascended the summit with Mr Low, reported that whenever they approached the spot where these diamonds showed themselves at a distance, they invariably disappeared: as these men have a perfect faith in every wild imagination of the *Arabian Nights*, they easily convinced themselves and their auditors that the jinn [spirits] would not permit them to take them. The old story of the great diamond, guarded on the summit of Kina Balu by a ferocious dragon, arose probably from some such cause. The Malays are great storytellers, and these wonders interest them. I may notice that most of the men that were with us accompanied us to the mountain of Molu [Mulu] the preceding February, and then one of the Borneans commenced a story which lasted the seventeen days we were away, and he occasionally went on with it during our present journey. It was the history of an unfortunate princess, who for 'seven days and seven nights neither eat nor drank, but only wept'.

Opposite our resting-place we observed some remarkably elegant tree ferns, whose stems rose occasionally to the height of ten feet, and with their long leaves bending gracefully on every side, they were an ornament to the river's bank. We noticed as yet but little old forest. The only fine trees we saw were near the villages, and these were preserved for their fruits. Where the land is not cultivated, it is either covered with brushwood, or trees of a young growth.

Drizzling rain prevented our departure till near eight, when we continued our course along the rice-fields: we had been told we should find the path very bad, but were agreeably surprised by it proving dry and principally among plantations of kiladi [arum]. We crossed the river only five times, and passed over a sandstone range about five hundred feet above the plain: it was nearly three miles from our resting-place. The stream had now become a perfect mountain torrent, breaking continually over rocks.

Occasionally the fords were difficult, as the continued rains rendered the river very full. At one place where an island divides the Tampasuk, it was so deep that it was found necessary to swim over, and only a very expert man could have done it, as the water rushed down with great force. The Bajus, however, were quite prepared; they did not attempt to cross the stream in a direct course, but allowed themselves to be carried away a little, and reached the other side about fifty yards farther down. They did it very cleverly, carrying all our luggage over, little by little, swimming with one hand and holding the baskets in the air with the other. As we could not swim, two men placed themselves, one on either side of us, told us to throw ourselves flat on the water and remain passive; in a few minutes we were comfortably landed on the opposite bank, drenched to the skin, it is true, but we had scarcely had any dry clothes on us during the whole journey; however, no sooner did we arrive at our resting-places, than we stripped, bathed, rubbed ourselves into a glow, and put on dry clothes. Nothing is so essential as this precaution, and I have twice had severe attacks of fever from neglecting it. The hills as we advanced began closing in on the river's banks, leaving occasionally but a narrow strip of flat ground near the stream.

At 11.20 A.M. we reached Koung, a large, scattered village on a grassy plain: it is a very pretty spot, the greensward extending to the river's banks, where the cattle and buffaloes graze: about a hundred feet up the side of a neighbouring hill is another portion of the village. The roaring torrent foams around, affording delicious spots for bathing, the water being delightfully cool. In the bed of the stream there were masses of angular granite, mixed with the water-worn boulders. It was the first time we had ever seen it of that sharp form, but similar blocks were afterwards noticed on the summit. The wild raspberry is very plentiful here. One cannot help having one's attention continually drawn to the air of comfort, or, rather, to the appearance of native wealth observed among the Ida'an: food in abundance, with cattle, pigs, fowls, rice, and vegetables; and no one near them to plunder or exact. Accustomed as I had been to the aborigines around the capital, the contrast struck me forcibly.

Next day we hoped to reach Kiau, the village from which Mr Low started for the mountain in the spring of 1851. There was an apparent hitch about getting from that place; but we thought perhaps the reports arose from tribal jealousy. At four P.M., Koung: barometer, 28.678°; thermometer, 77.5°; unattached, 78.3°. So that this village must be about 1,500 feet above the level of the sea: a very rapid rise for the stream in so short a distance. The sandstone hill we crossed to-day had the same characteristics as those I had observed up the Sakarang, Batang Lupar, and near the capital—all being very steep, with narrow ridges, and buttresses occasionally springing from their sides: on the one we crossed to-day was a quantity of red shale.

Near our last night's resting-place, I noticed, for the first time on this river, some sago palms; they have again shown themselves to-day, and there are a few round the village, but neither these trees nor cocoanut nor areca palms are plentiful. At every village I made inquiries about cotton, and, like the men with tails, it was always grown a little farther off; only we know cotton must be grown somewhere in this neighbourhood, as at the very moment I was writing my journal I saw an old woman engaged spinning yarn from native material. The Lanuns [Illanun] also furnish a cloth which is highly prized among every class of inhabitants in Borneo; it is a sort of checked black cloth, with narrow lines of white running through it, and glazed on one [side]. This was formerly made entirely of native yarn; but I am afraid this industry will soon decline, as connoisseurs are already beginning to discover that the Lanun women, finding English yarn so cheap, are using it in preference, though it renders the article much less durable. It is also worthy of notice that this cloth is dyed from indigo grown on the spot. These Ida'an purchase their supplies of cotton off the Inserban and Tuhan Ida'an who live on the road to the lake, while the Bajus obtain theirs from the Lobas near Maludu [Marudu] Bay. I saw one plant growing near the hut where we rested last night; it was about ten feet high, and covered with flowers.

They told us at Koung that the Ida'an were at war; but though they may have quarrels, they must be trifling, as we met every day women and children by themselves at considerable distances from their houses. Besides, parties of a dozen men and boys of the supposed enemies passed us on their way to Tampasuk to trade, and in none of their villages did we notice heads.

All these Ida'an appear to pay particular attention to the cultivation of the Kiladi (arum), planting it in their fields immediately after gathering in the rice crop, and keeping it well weeded: they grow it everywhere, and it must afford them abundance of food. It is in shape something like a beet-root, and has the flavour of a yam. Roasted in the ashes, and brought smoking hot to table, torn open, and adding a little butter, pepper, and salt, it is very palatable, particularly among those hills.

Saduk bore N. E. and Kina Balu due E. from the southern portion of the village.

Started about seven in a S. E. by E. direction, ascending a hill on which the village of Labang Labang is situated: here occurred a scene. Mr Low and I, with a few men, were walking ahead of the party: as we passed the first house, an old woman came to the door, and uttered some sentences which struck us as sounding like a curse: however, we took no notice; but as we approached the end of the village, we were hailed by an ugly-looking fellow, with an awful squint, who told us to stop, as we should not pass through his village: this was evidently a prepared scene, the whole of the population turning out, armed: so we did stop to discuss the point. We asked what he meant: he answered that they had never had good crops since Mr Low ascended the mountain in 1851, and gave many other sapient reasons why we should not ascend it now; but he wound up by saying that if we would pay a slave as black mail, they would give us permission to pass and do as we pleased: this showed us that nothing but extortion was intended; yet, to avoid any disagreeable discussion, we offered to make him a present of forty yards of grey shirting; but this proposition was not listened to, and he and his people became very insolent in their manner.

We sent back one of the men to hurry up the stragglers, and in the meantime continued the discussion. They then said they would take us up the mountain if we would start from their village; but being unwilling to risk a disappointment, we declined. They remembered how the Kiaus had turned back Mr Lobb, because he would not submit to their extortions, and thought they might do the same with us. As the Ida'an were shaking their spears and giving other hostile signs, we thought it time to bring this affair to a climax; so I ordered the men to load their muskets, and Mr Low, stepping up to the chief with his five-barrelled pistol, told the interpreter to explain that we were peaceable travellers, most unwilling to enter into any contest; that we had obtained the permission of the Government of the country, and that we were determined to proceed; that if they carried out their threats of violence, he would shoot five with his revolver, and that I was prepared to do the same with mine; that they might, by superior numbers, overcome us at last, but in the meantime we would make a desperate fight of it.

This closed the scene: as long as we had only half a dozen with us, they were bullies; but as our forces began to arrive, and at last amounted to fifty men, with twenty musket-barrels shining among them, they became as gentle as lambs, and said they would take two pieces of grey shirting; but we refused to give way, keeping to our original offer, and then only if the chief would follow us on our return, and receive it at Tampasuk. We ordered the men to advance, and we would close up the rear: no opposition was offered; on the contrary, the chief accompanied us on our road, and we had no more trouble with the Labang Labang people. We were detained forty minutes by this affair. Our guides explained the matter to us: when Mr Low was here last time, many reports were spread of the riches which the Kiaus had obtained from the white man, and they were jealous that the other branch of their tribe should obtain the wealth that was passing from them through their village. The Koung people tried to persuade us last night to start from their place, and as they were very civil we should have liked to oblige them, but they were uncertain whether they could take us to the summit. Mr Lobb, when he reached Kiau, had but a small party, and was unarmed, so they would not allow him to pass, except on terms that were totally inadmissible.

Immediately after passing the village, we descended a steep and slippery path to one of the torrents into which the Tampasuk now divided. After crossing it, we were at the base of the spur on which the village of Kiau is situated. We passed several purling streams which descended, in a winding course, the face of the hill. From one spot in our walk, we had a beautiful view of two valleys, cultivated on both banks, with the foaming streams dashing among the rocks below. Over the landscape were scattered huts, which had the peculiarity of being flat-roofed: the Kiaus using the bamboo as the Chinese use their tiles, split in two; the canes are arranged side by side across the whole roof, with their concave sides upwards to catch the rain; then a row placed convex to cover the edges of the others, and prevent the water dropping through. They are quite watertight, and afforded an excellent hint for travellers where bamboos abound.

The latter portion of the road was difficult climbing, the clay being slippery from last night's rain; but as we approached our resting-place, the walking became easier. Kiau is a large village on the southern side of the spur. The houses scattered on its face are prettily concealed from each other by clumps of cocoanuts and bamboos. It covers a great extent of ground, but is badly placed, being more than 800 feet above the torrent—that is, the portion of the village at which we stayed. The eastern end was nearer the stream. The inhabitants supplied themselves with drinking-water from small rills which were led in bamboos to most of their doors. We brought up about eleven, our course being generally E. S. E. Thermometer 73° at twelve in the house. We felt it chilly, and took to warm clothing.

The Kiaus are much dirtier than any tribes I have seen in the neighbourhood: the children and women are unwashed, and most of them are troubled with colds, rendering them in every sense unpleasant neighbours. In fact, to use the words of an experienced traveller, 'they cannot afford to be clean', their climate is chilly, and they have no suitable clothing. We observed that the features of many of these people were very like Chinese—perhaps a trace of that ancient kingdom of Celestials that tradition fixes to this neighbourhood. They all showed the greatest and most childlike curiosity at everything either we or our servants did.

In the afternoon, Lemaing, Mr Low's old guide, came in. Mr Low recognized his voice immediately, though seven years had passed since he had heard it. Sir James Brooke has a most extraordinary faculty of remembering voices, as well as names, even of natives whom he has only seen once. It is very useful out here, and I have often found the awkwardness which arises from my quickly forgetting both voices and names.

Shortly after Lemaing's arrival, a dispute arose between him and Lemoung, the chief of the house in which we were resting: both voices grew excited; at last, they jumped up, and each spat upon the floor in a paroxysm of mutual defiance: here we interposed to preserve the peace, and calm being restored, it was found that seven years ago they had disputed about the division of Mr Low's goods, and the quarrel had continued ever since—the whole amount being five dollars. Lemoung said that his house had been burnt down in consequence of the white man ascending Kina Balu, and that no good crops of rice had grown since; but it was all envy; he thought in the distribution he had not secured a fair share. We asked if he had ascended the mountain; he said no, but his son had brought some rice, for which, on inquiry, we found he had been paid. Drizzling rain the whole afternoon.

The thermometer registered 66° last night, and we enjoyed our sleep under blankets. At mid-day, we took out the barometer from its case, and found, to our inexpressible vexation, that it was utterly smashed. This will destroy half the pleasure of the ascent; in fact, our spirits are somewhat depressed by the accident, and by Mr Low's feet getting worse. At twelve, thermometer 77°. (The lamentable accident so disgusted me that I find no further entry in that day's journal, but a pencilled note remarks that the Ida'an preserve their rice in old bamboos two fathoms long, which are placed on one side of the doorway. It is said that these bamboos are preserved for generations, and, in fact, they looked exceedingly ancient.)

Last night, thermometer 69°. At early dawn, we heard the war-drums beating in several houses, and shouts and yells from the boys. They said it was a fête day, but we rightly guessed it had something to do with our expedition. For some time, our guide did not make his appearance, and a few young fellows on the hill over the village threw stones as we appeared at the door—a very harmless demonstration, as they were several hundred yards off—but discharging and cleaning a revolver lessened the amount of hostile shouting. About nine, the guide made his appearance; the women seemed to enjoy the scene, and followed us to witness the skirmish; but the enemy, if there were an enemy, did not show, and the promised ambush came to nothing—it was but a trick of Lemoung to try and disgust Lemaing, and frighten us by the beating of drums and shouting. At the place where we were assured an attack would be made, we found but a few harmless women carrying tobacco.

Our path lay along the side of the hill in which the village stands, we followed it about four miles in an easterly direction, and then descended to a torrent, one of the feeders of the Tampasuk, where we determined to spend the night, as Mr Low's feet were becoming very swollen and painful, and it was as well to collect the party. We had passed through considerable fields of sweet potatoes, kiladi, and tobacco, where the path was crossed occasionally by cool rills from the mountains. We enjoyed the cold water very much, and had a delightful bath. The torrent comes tumbling down, and forms many fine cascades. Mr Low botanized a little, notwithstanding his feet were suppurating. The hut in which we spent the night was very pretty-looking, flat-roofed, built entirely of bamboos.

To-day, we had a specimen of the thieving of our Ida'an followers. One man was caught burying a tin of sardines; another stole a Bologna sausage, for which, when hungry, I remembered him, and another a fowl.

Next morning, Mr Low found it impossible to walk, and I was therefore obliged to start without him. We showed our perfect confidence in the villagers of Kiau by dividing our party, leaving only four men with Mr Low to take care of the arms; we carried with us up the mountain nothing but our swords and one revolver. They must have thought us a most extraordinary people; but we knew that their demonstrations of hostility were really harmless, and more aimed against each other than against us. Probably, had we appeared afraid, it might have been a different matter.

Our course was at first nearly east up the sub-spur of a great buttress. The walking was severe, from the constant and abrupt ascents and descents, and the narrowness of the path when it ran along the sides of the hill, where it was but the breadth of the foot. At one place we had a view of a magnificent cascade. The stream that runs by the cave, which is to be one of our resting-places, falls over the rocks forming minor cascades; then coming to the edge of the precipice, throws itself over, and in its descent of above fifteen hundred feet appears to diffuse itself in foam, ere it is lost in the depths of the dark-wooded ravines below.

I soon found I had made a great mistake in permitting these active mountaineers to lead the way at their own pace, as before twelve o'clock I was left alone with them, all my men being far behind, as they were totally unaccustomed to the work. Arriving at a little foaming rivulet, I sat down and waited for the rest of the party, and when they came up, they appeared so exhausted that I had compassion on them, and agreed to spend the night here. The Ida'an were very dissatisfied, and declared they would not accompany us, if we intended to make such short journeys; but we assured them that we would go on alone if they left us, and not pay them the stipulated price for leading us to the summit. I soon set the men to work to build a hut of long poles, over which we could stretch our oiled cloths, and to make a raised floor to secure us from being wet through by the damp moss and heavy rain that would surely fall during the night. At three P.M. the thermometer fell to 65°, which to the children of the plain rendered the air unpleasantly cold; but we worked hard to collect boughs and leaves to make our beds soft; and wood was eagerly sought for to make fires in the holes beneath our raised floor. This filled the place with smoke, but gave some warmth to the men.

The Ida'an again tried to get back, but I would not receive their excuse that they would be up early in the morning: they then set hard at work going through incantations to drive away sickness. The guide Lemaing carried an enormous bundle of charms, and on him fell the duty of praying or repeating some forms: he was at it two hours by my watch. To discover what he said, or the real object to whom he addressed himself, was almost impossible through the medium of our bad interpreters. I could hear him repeating my name, and they said he was soliciting the spirits of the mountain to favour us.

The thermometer registered 57° last night in tent. Started at seven; I observed a fine yellow sweet-scented rhododendron on a decayed tree, and requested my men on their return to take it to Mr Low; continuing the ascent, after an hour's tough walking, reached the top of the ridge. There it was better for a short time; but the forest, heavily hung with moss, is exposed to the full force of the south-west monsoon, and the trees are bent across the path, leaving occasionally only sufficient space to crawl through. We soon came upon the magnificent pitcher-plant, the *Nepenthes Lowii*, that Mr Low was anxious to get. We could find no young plants, but took cuttings, which the natives said would grow.

We stopped to breakfast at a little swampy spot, where the trees are becoming very stunted, though in positions protected from the winds they grow to a great height. Continuing our course, we came upon a jungle that appeared to be composed almost entirely of rhododendrons, some with beautiful pink, crimson, and yellow flowers. I sat near one for about half an hour apparently in intense admiration, but, in fact, very tired, and breathless, and anxious about my followers, only one of whom had kept up with me.

Finding it useless to wait longer, as the mist was beginning to roll down from the summit, and the white plain of clouds below appeared rising, I pushed on to the cave, which we

121

intended to occupy. It was a huge granite boulder, resting on the hill side, that sheltered us but imperfectly from the cold wind. The Ida'an, during the day, amused themselves in trying to secure some small twittering birds, which looked like canaries, with a green tint on the edges of their wings, but were unsuccessful. They shot innumerable pellets from their blowpipes, but did not secure one. In fact, they did not appear to use this instrument with any skill.

At four o'clock the temperature of the air was 52°, and of the water 48°.

Some of my men did not reach us till after dark, and it was with great difficulty that I could induce the Malays to exert themselves to erect the oiled clothes, to close the mouth of the cave, and procure sufficient firewood. They appeared paralyzed by the cold, and were unwilling to move.

During the night, the thermometer at the entrance of the cave fell to 36° 5'; and on my going out to have a look at the night-scene, all the bushes and trees appeared fringed with hoar frost.

After breakfasting at the cave, we started for the summit. Our course lay at first through a thick low jungle, full of rhododendrons; it then changed into a stunted brushwood, that almost hid the rarely-used path; gradually the shrubs gave way to rocks, and then we commenced our ascent over the naked granite. A glance upwards from the spot where we first left the jungle, reveals a striking scene—a face of granite sweeping steeply up for above 3,000 feet to a rugged edge of pointed rocks; while on the farthest left the southern peak looked from this view a rounded mass. Here and there small runnels of water passed over the granite surface, and patches of brushwood occupied the sheltered nooks. The rocks were often at an angle of nearly forty degrees, so that I was forced to ascend them, at first, with woollen socks, and when they were worn through, with bare feet. It was a sad alternative, as the rough stone wore away the skin and left a bleeding and tender surface.

After hard work, we reached the spot where Mr Low had left a bottle, and found it intact—the writing in it was not read, as I returned it unopened to its resting-place.

Low's Gully is one of the most singular spots in the summit. We ascend an abrupt ravine, with towering perpendicular rocks on either side, till a rough natural wall bars the way. Climbing on this, you look over a deep chasm, surrounded on three sides by precipices, so deep that the eye could not reach the bottom; but the twitter of innumerable swallows could be distinctly heard, as they flew in flocks below. There was no descending here: it was a sheer precipice of several thousand feet, and this was the deep fissure pointed out to me by Mr Low from the cocoanut grove on the banks of the Tampasuk when we were reclining there, and proved that he had remembered the very spot where he had left the bottle.

I was now anxious to reach one of those peaks which are visible from the sea; so we descended Low's Gully, through a thicket of rhododendrons, bearing a beautiful blood-coloured flower, and made our way to the westward. It was rough walking at first, while we continued to skirt the rocky ridge that rose to our right; but gradually leaving this, we advanced up an incline composed entirely of immense slabs of granite, and reaching the top, found a noble terrace, half a mile in length, whose sides sloped at an angle of thirty degrees on either side. The ends were the Southern Peak and a huge cyclopean wall.

I followed the guides to the former, and after a slippery ascent, reached the summit. I have mentioned that this peak has a rounded aspect when viewed from the eastward; but from the northward it appears to rise sharply to a point; and when with great circumspection I crawled up, I found myself on a granite point, not three feet in width, with but a water-worn way a few inches broad to rest on, and prevent my slipping over the sloping edges.

During the climbing to-day, I suffered slightly from shortness of breath, and felt some disinclination to bodily exertion; but as soon as I sat down on this lofty point, it left me, and a feeling came on as if the air rendered me buoyant and made me long to float away.

Calmly seated here, I first turned my attention to the other peaks which stretched in a curved line from east to west, and was rather mortified to find that the most westerly and another to the east appeared higher than where I sat, but certainly not more than a hundred feet. The guides called this the mother of the mountain, but her children may have outgrown her. Turning to the south-west, I could but obtain glimpses of the country, as many thousand feet below masses of clouds passed continually over the scene, giving us but a partial view of sea, and rivers, and hills. One thing immediately drew my attention, and that was a very lofty peak towering above the clouds, bearing S. $\frac{1}{2}$ E. It appeared to be an immense distance off, and I thought it might be the great mountain of Lawi, of which I went in search some months later; but it must be one much farther to the eastward, and may be the summit of Tilong, which, as I have before mentioned, some declare to be much more lofty than Balu itself.

Immediately below me, the granite for a thousand feet sloped sharply down to the edge of that lofty precipice that faces the valley of Pinokok to the south-west. I felt a little nervous while we were passing along this to reach the southern peak, as on Mr Low's former expedition a Malay had slipped at a less formidable spot, and been hurried down the steep incline at a pace that prevented any hope of his arresting his own progress, when leaning on his side his kris fortunately entered a slight cleft, and arrested him on the verge of a precipice.

Among the detached rocks and in the crevices grew a kind of moss, on which the Ida'an guides declared the spirits of their ancestors fed. A grass also was pointed out that served for the support of the ghostly buffaloes which always followed their masters to the other world. As a proof, the print of a foot was shown me as that of a young buffalo; it was not very distinct, but appeared more like the impression left by a goat or deer.

Our guides became very nervous as the clouds rose and now occasionally topped the precipice, and broke, and swept up the slopes, enveloping us. They urged me to return; I saw it was necessary, and complied, as the wind was rising, and the path we were to follow was hidden in mist.

We found the air pleasantly warm and very invigorating; the thermometer marked 62° in the shade; and as we perceived little rills of water oozing from among the granite rocks, the summit would prove a much better encamping ground than our cold cave, where the sun never penetrates. The Ida'an, however, feared to spend one night in this abode of spirits, and declined carrying my luggage.

Our return was rather difficult, as the misty rain rendered the rocks slippery, but we all reached the cave in safety. Here

I received a note from Mr Low, but he was still unable to walk. The bathing water was 49°.

During the night the temperature fell, and the registering thermometer marked 41°. My feet were so injured by yesterday's walking that I was unable to reascend the mountain to collect plants and flowers, so sent my head man Musa with a large party. I, however, strolled about a little to look for seeds and a sunny spot, as the ravine in which our temporary home was, chilled me through. I was continually enveloped in mist, and heard afterwards to my regret that the summit was clear, and that all the surrounding country lay exposed to view. The low, tangled jungle was too thick to admit of our seeing much. I climbed the strongest and highest trees there, but could only get glimpses of distant hills.

Thermometer during the night, 43°, while in the cave yesterday it marked 56° at two o'clock.

Started early to commence our descent, collecting a few plants on our way; the first part of the walking is tolerably good—in fact, as far as the spot where we rested for breakfast on our ascent. It is in appearance a series of mighty steps. Passed on the wayside innumerable specimens of that curious pitcher-plant the *Nepenthes villosa*, with serrated lips.

After leaving the great steps, our course was along the edge of a ridge, where the path is extremely narrow; in fact, in two or three places not above eighteen inches wide—a foot of it serving as parapet, six inches of sloping rock forming the path. From one of these craggy spots a noble landscape is spread before us, eighty miles of coast-line, with all the intervening country being visible at once. With one or two exceptions, plains skirt the sea-shore, then an undulating country, gradually rising to ranges varying from two to three thousand feet, with glimpses of silvery streams flowing among them. The waters of the Mengkabong and Sulaman, swelling to the proportions of lakes, add a diversity to the scene.

It is fortunate that the ridge is not often so narrow as at these spots; for on one side there is a sheer descent of fifteen hundred feet, and on the other is very perpendicular ground, but wooded. Two decaying rocks that obstruct the path are also dangerous to pass, as we had to round them, with uncertain footing, and nothing but a bare, crumbling surface to grasp. With the exception of these, the path is not difficult or tiring, until we leave the ridge and descend to the right towards the valleys: then it is steep, slippery, and very fatiguing, and this continues for several miles, until we have lowered the level nearly four thousand feet. The path, in fact, is as vile as path can be.

By the time I reached the hut where I had left Mr Low, I felt completely exhausted; but a little rest, a glass of brandy-and-water, and a bathe in the dashing torrent that foamed among the rocks at our feet, thoroughly restored me. The water here felt pleasant after the bitter cold of that near the cave. My companion had employed his time collecting plants, though his feet were not at all better.

Next morning we manufactured a kind of litter, on which Mr Low was to be carried, and then started along a path that skirted the banks of the Kalupis, that flows beneath the village, and is, in fact, the source of the Tampasuk. We passed through several fields of tobacco, as well as of yams and kiladis; the first is carefully cultivated, and not a weed was to be observed among the plants. Leaving the water, we pushed up the steep bank to the lower houses of the village, and made our way on to Lemoung's, to reach it just as a drenching shower came on.

Here we found one of our Baju guides, who had been sent back to construct rafts for the return voyage. I was not sorry to find that some had been prepared, as it appeared otherwise necessary that Mr Low should be carried the whole way.

The villagers said they were at war even during the time we were at their houses with a neighbouring tribe, which induces them to bear arms wherever they may go; but the whole affair must be very trifling, as they sleep at their farms, and we saw, totally unprotected, troops of girls and women at work in the fields.

We thought it better to make some complaints of the dishonesty shown, before we ascended the mountain; they were profuse in apologies, but they had evidently enjoyed the sausage.

We spent the afternoon and evening in settling all claims against us, and having completed that work, ordered the rest of our baggage to be packed up ready for an early start next morning. Among the undistributed goods was about twenty pounds weight of thick brass wire. While I was away bathing, Lemaing coolly walked off with it; but on my return Mr Low informed me of what had occurred. Knowing that if we permitted this to pass unnoticed, it would be a signal for a general plunder, we determined to recover the wire. As Mr Low could not move, I went by myself in search of Lemaing, and soon heard his voice speaking loudly in the centre of a dense crowd of the villagers. I forced my way through, and found him seated, with the brass wire in his hand, evidently pointing out its beauty to an admiring audience. I am afraid I very much disconcerted him, as with one hand I tore the prize from his grasp, and with the other put a revolver to his head, and told him to beware of meddling with our baggage. I never saw a look of greater astonishment; he tried to speak, but the words would not come, and the crowd opening, I bore back the trophy to our end of the village house.

The Bajus told us we should find the Ida'an of the plains dishonest, while those of the hills had the contrary reputation. We lost nothing in the plains; here we had to guard carefully against pilferers.

We noticed that as we gradually receded from the sea, the clothing of the inhabitants became less—on the plains all the Ida'an wore trousers and jackets; at Koung and Kiau very few, and we were assured that those in the interior wore nothing but bark waist-cloths.

An incident occurred the evening before our departure, which showed how the Ida'an distrust each other. Among the goods we paid to our guides were twenty fathoms of thick brass wire; the coils were put down before them; they talked over it for two hours, and could not settle either the division, or who should take care of it until morning; at length one by one all retired and left the wire before us, the last man pushing it towards Musa, asking him to take charge of it. Not relishing this trust, he carried it to Li Moung's house, and placing it in the midst of the crowd, left it, and they then quarrelled over it till morning.

We thought last night every claim had been settled, but this morning they commenced again, anxious to prevent any goods leaving their village. We ourselves did not care to take back to our pinnace anything that was not necessary to enable us to pay our way. We made liberal offers to them if they would carry Mr Low to the next village, but they positively refused to assist us farther. We therefore collected our Malays outside the place, and prepared to start; and were on the point of doing so, when shouts in the village house attracted our attention, and a man ran out to say that they were plundering the baggage left in charge of the Bungol Ida'an. As this consisted of our clothes and cooking utensils, it was not to be borne, and I ran back into the house, where I found a couple of hundred men surrounding our Ida'an followers and undoing the packages; they were startled by the sight of my rifle, and when they heard the rush caused by the advance of Mr Low and our Malays, they fled to the end of the house, and soon disappeared through the opposite door. The panic seemed to cause the greatest amusement to the girls of this house, who talked and laughed, and patted us on the shoulders, and appeared to delight in the rapid flight of their countrymen. None of their own relatives, however, had joined in the affair.

Mr Low's rapid advance to my support surprised me; but I found that with the assistance of a servant he had hopped the whole of the way, revolver in hand. Our men behaved with remarkable resolution, and would have driven off the whole village had it been necessary. One Malay got so excited, that he commenced a war dance, and had we not instantly interfered, would have worked himself up to run a muck among the Ida'an. Though we wished to frighten them into honesty towards us, we were most anxious that not the slightest wound should be given, and I may here remark, that in none of our journeys have we ever found it necessary to use our weapons against the inhabitants. We discovered that showing ourselves prepared to fight, if necessary, prevented its being ever necessary to fight.

We pushed on to Koung by a path that led below Labang Labang, Mr Low suffering severely from the necessity of having to walk six miles over stony country with suppurating feet.

At Koung we vainly endeavoured to obtain a buffalo, on which Mr Low might ride; but the villagers showed no inclination to assist. So next morning we pushed on through heavy rain to the village of Tambatuan, where the Tampasuk becomes a little more fit for rafts. I was glad to see Mr Low safely there, and then, as the rafts would not hold us all, I walked on with the men. The heavy rain had caused the river to swell, and the walking and the fording were doubly difficult, but we continued our course, and in two days reached the village of Ginambur, and joining Mr Low on the raft, pursued our journey to the Datu's house.

Next day to the Abai; but contrary winds prevented our reaching Labuan for five days.

We were not quite satisfied with the results of this expedition, and determined to start again, but choosing another route, the same followed by Mr Low in 1851.

Life in the Forests of the Far East, 2 vols; Smith, Elder and Co., London, 1862, pp. 231, 251–79.

9
A Journey among the Land Dayaks (Bidayuhs) of Sarawak

ALFRED RUSSEL WALLACE

One of the world's most distinguished naturalists, Alfred Wallace spent eight years between 1854 and 1862 travelling and collecting in the East Indies. Prior to his Eastern expedition he had already participated in a scientific exploration of Amazonia, in 1848–50. The collections which he made in the Malay archipelago for himself, museums, and interested amateur naturalists were truly extensive. He amassed over 125,000 specimens, including beetles, butterflies, bird skins, quadrupeds, and land shells. He tells us that the long delay between his return to England in 1862 and the publication of his book in 1869 was due primarily to the formidable task of sorting, arranging, and working on his collections, but he had also been in a weakened state of health following his arduous travels in the tropics. He had covered altogether 14,000 miles in the archipelago, comprising 60–70 separate journeys. His was not a large scientific expedition. Although it was officially sponsored by the British and Dutch governments, Wallace travelled as a private individual. He only employed one or two, sometimes three, servants at a time to assist him. For nearly half his time abroad he was accompanied by his assistant Charles Allen, who was with him in Sarawak.

Wallace arrived in Sarawak on 1 November 1854 and left from there for Singapore on 25 January 1856, having spent over a year in Borneo on the first stage of his travels. While in Sarawak Sir James Brooke, who himself took a keen interest in natural history, offered Wallace every assistance. Wallace also engaged a local Malay servant, Ali.

As John Bastin has said of Wallace's text: 'This is the most famous of all books on the Malay Archipelago. It is the greatest travel book on the region and, in its analysis of the geographical distribution of animals, it ranks as one of the most important books of the nineteenth century' (1986: vii).[1] The Malay Archipelago is an absorbing book, although another apparent reason why it was long in the making was that Wallace disliked writing narrative and was terribly hesitant about the quality of his work (Bastin, p. viii). He need not have had any such self-doubts. The book became enormously popular.

The extract which follows recounts Wallace's journey into the interior of Sarawak between November 1855 and January 1856. In contrast to Beccari, Wallace was 'as much interested in man as in the natural world' (Bastin, p. xxi), and this extract—Chapter V of his book—provides a delightful combination of human and natural details. Wallace tells us about the Land Dayak communities inland of Kuching, the capital of Sarawak, as well as about bamboo and the distinctive tropical fruit durian. Not to everyone's liking, Wallace obviously enjoyed durian, and his is, I think, the best succinct description we have of its texture and taste.

The inclusion of a part of Wallace's account of Sarawak has another significance. The Malay Archipelago *was dedicated to Charles Darwin, and it was during this stay in Sarawak that Wallace independently began to formulate the theory of the natural selection of species.*

A S the wet season was approaching I determined to return to Sarawak, sending all my collections with Charles Allen round by sea, while I myself proposed to go up to the sources of the Sadong River, and descend by the Sarawak valley. As the route was somewhat difficult, I

---

[1]John Bastin, 'Introduction' to *The Malay Archipelago. The Land of the Orang Utan, and the Bird of Paradise*, Oxford University Press reprint, Singapore, 1986, pp. vii–xxvii.

took the smallest quantity of baggage, and only one servant, a Malay lad named Bujon, who knew the language of the Sadong Dyaks, with whom he had traded. We left the mines on the 27th of November, and the next day reached the Malay village of Gudong, where I stayed a short time to buy fruit and eggs, and called upon the Datu Bandar, or Malay governor of the place. He lived in a large and well-built house, very dirty outside and in, and was very inquisitive about my business, and particularly about the coal mines. These puzzle the natives exceedingly, as they cannot understand the extensive and costly preparations for working coal, and cannot believe it is to be used only as fuel when wood is so abundant and so easily obtained. It was evident that Europeans seldom came here, for numbers of women skeltered away as I walked through the village; and one girl about ten or twelve years old, who had just brought a bamboo full of water from the river, threw it down with a cry of horror and alarm the moment she caught sight of me, turned round and jumped into the stream. She swam beautifully, and kept looking back as if expecting I would follow her, screaming violently all the time; while a number of men and boys were laughing at her ignorant terror.

At Jahi, the next village, the stream became so swift in consequence of a flood, that my heavy boat could make no way, and I was obliged to send it back and go on in a very small open one. So far the river had been very monotonous, the banks being cultivated as rice-fields, and little thatched huts alone breaking the unpicturesque line of muddy bank crowned with tall grasses, and backed by the top of the forest behind the cultivated ground. A few hours beyond Jahi we passed the limits of cultivation, and had the beautiful virgin forest coming down to the water's edge, with its palms and creepers, its noble trees, its ferns, and epiphytes. The banks of the river were, however, still generally flooded, and we had some difficulty in finding a dry spot to sleep on. Early in the morning we reached Empugnan, a small Malay village situated at the foot of an isolated mountain which had been visible from the mouth of the Simunjon River. Beyond here the tides are not felt and we now entered upon a district of elevated forest, with a finer vegetation. Large trees stretch out their arms across the stream, and the steep, earthy banks are clothed with ferns and Zingiberaceous plants.

Early in the afternoon we arrived at Tabokan, the first village of the Hill Dyaks. On an open space near the river about twenty boys were playing at a game something like what we call 'prisoner's base'; their ornaments of beads and brass wire and their gay-coloured kerchiefs and waist-cloths showing to much advantage, and forming a very pleasing sight. On being called by Bujon, they immediately left their game to carry my things up to the 'head-house',—a circular building attached to most Dyak villages, and serving as a lodging for strangers, the place for trade, the sleeping-room of the unmarried youths, and the general council-chamber. It is elevated on lofty posts, has a large fireplace in the middle and windows in the roof all round, and forms a very pleasant and comfortable abode. In the evening it was crowded with young men and boys, who came to look at me. They were mostly fine young fellows, and I could not help admiring the simplicity and elegance of their costume. Their only dress is the long 'chawat', or waist-cloth, which hangs down before and behind. It is generally of blue cotton, ending in three broad bands of red, blue, and white. Those who can afford it wear a handkerchief on the head, which is either red, with a narrow border of gold lace, or of three colours, like the 'chawat'. The large flat moon-shaped brass earrings, the heavy necklace of white or black beads, rows of brass rings on the arms and legs, and armlets of white shell, all serve to relieve and set off the pure reddish brown skin and jet-black hair. Add to this the little pouch containing materials for betel-chewing and a long slender knife, both invariably worn at the side, and you have the everyday dress of the young Dyak gentleman.

The 'Orang Kaya', or rich man, as the chief of the tribe is called, now came in with several of the older men; and the 'bitchara' or talk commenced, about getting a boat and men to take me on the next morning. As I could not understand a word of their language, which is very different from Malay, I took no part in the proceedings, but was represented by my boy Bujon, who translated to me most of what was said. A Chinese trader was in the house, and he, too, wanted men the next day; but on his hinting this to the Orang Kaya, he was sternly told that a white man's business was now being discussed, and he must wait another day before his could be thought about.

After the 'bitchara' was over and the old chiefs gone, I

asked the young men to play or dance, or amuse themselves in their accustomed way; and after some little hesitation they agreed to do so. They first had a trial of strength, two boys sitting opposite each other, foot being placed against foot, and a stout stick grasped by both their hands. Each then tried to throw himself back, so as to raise his adversary up from the ground, either by main strength or by a sudden effort. Then one of the men would try his strength against two or three of the boys; and afterwards they each grasped their own ankle with a hand, and while one stood as firm as he could, the other swung himself round on one leg, so as to strike the other's free leg, and try to overthrow him. When these games had been played all round with varying success, we had a novel kind of concert. Some placed a leg across the knee, and struck the fingers sharply on the ankle, others beat their arms against their sides like a cock when he is going to crow, thus making a great variety of clapping sounds, while another with his hand under his armpit produced a deep trumpet note; and, as they all kept time very well, the effect was by no means unpleasing. This seemed quite a favourite amusement with them, and they kept it up with much spirit.

The next morning we started in a boat about thirty feet long, and only twenty-eight inches wide. The stream here suddenly changes its character. Hitherto, though swift, it had been deep and smooth, and confined by steep banks. Now it rushed and rippled over a pebbly, sandy, or rocky bed, occasionally forming miniature cascades and rapids, and throwing up on one side or the other broad banks of finely coloured pebbles. No paddling could make way here, but the Dyaks with bamboo poles propelled us along with great dexterity and swiftness, never losing their balance in such a narrow and unsteady vessel, though standing up and exerting all their force. It was a brilliant day, and the cheerful exertions of the men, the rushing of the sparkling waters, with the bright and varied foliage which from either bank stretched over our heads, produced an exhilarating sensation which recalled my canoe voyages on the grander waters of South America.

Early in the afternoon we reached the village of Borotoi, and, though it would have been easy to reach the next one before night, I was obliged to stay, as my men wanted to return and others could not possibly go on with me without the preliminary talking. Besides, a white man was too great a rarity to be allowed to escape them, and their wives would never have forgiven them if, when they returned from the fields, they found that such a curiosity had not been kept for them to see. On entering the house to which I was invited, a crowd of sixty or seventy men, women, and children gathered round me, and I sat for half an hour like some strange animal submitted for the first time to the gaze of an inquiring public. Brass rings were here in the greatest profusion, many of the women having their arms completely covered with them, as well as their legs from the ankle to the knee. Round the waist they wear a dozen or more coils of fine rattan stained red, to which the petticoat is attached. Below this are generally a number of coils of brass wire, a girdle of small silver coins, and sometimes a broad belt of brass ring armour. On their heads they wear a conical hat without a crown, formed of variously coloured beads, kept in shape by rings of rattan, and forming a fantastic but not un-picturesque head-dress.

Walking out to a small hill near the village, cultivated as a rice-field, I had a fine view of the country, which was becoming quite hilly, and towards the south, mountainous. I took bearings and sketches of all that was visible, an operation which caused much astonishment to the Dyaks who accompanied me, and produced a request to exhibit the compass when I returned. I was then surrounded by a larger crowd than before, and when I took my evening meal in the midst of a circle of about a hundred spectators anxiously observing every movement and criticising every mouthful, my thoughts involuntarily recurred to the lions at feeding time. Like those noble animals, I too was used to it, and it did not affect my appetite. The children here were more shy than at Tabokan, and I could not persuade them to play. I therefore turned showman myself, and exhibited the shadow of a dog's head eating, which pleased them so much that all the village in succession came out to see it. The 'rabbit on the wall' does not do in Borneo, as there is no animal it resembles. The boys had tops shaped something like whipping-tops, but spun with a string.

The next morning we proceeded as before, but the river had become so rapid and shallow, and the boats were all so small, that though I had nothing with me but a change of clothes, a gun, and a few cooking utensils, two were required to take me on. The rock which appeared here and there on the river-bank was an indurated clay-slate, sometimes crystalline, and thrown up almost vertically. Right and left of us rose isolated limestone mountains, their white precipices glistening in the sun and contrasting beautifully with the luxuriant vegetation that elsewhere clothed them. The river bed was a mass of pebbles, mostly pure white quartz, but with abundance of jaspar and agate, presenting a beautifully variegated appearance. It was only ten in the morning when we arrived at Budw, and, though there were plenty of people about, I could not induce them to allow me to go on to the next village. The Orang Kaya said that if I insisted on having men of course he would get them; but when I took him at his word and said I must have them, there came a fresh remonstrance; and the idea of my going on that day seemed so painful that I was obliged to submit. I therefore walked out over the rice-fields, which are here very extensive, covering a number of the little hills and valleys into which the whole country seems broken up, and obtained a fine view of hills and mountains in every direction.

In the evening the Orang Kaya came in full dress (a spangled velvet jacket, but no trousers), and invited me over to his house, where he gave me a seat of honour under a canopy of white calico and coloured handkerchiefs. The great verandah was crowded with people, and large plates of rice with cooked and fresh eggs were placed on the ground as presents for me. A very old man then dressed himself in bright-coloured cloths and many ornaments, and sitting at the door, murmured a long prayer or invocation, sprinkling rice from a basin he held in his hand, while several large gongs were loudly beaten and a salute of muskets fired off. A large jar of rice wine, very sour but with an agreeable flavour, was then handed round, and I asked to see some of their dances. These were, like most savage performances, very dull and ungraceful affairs; the men dressing themselves absurdly like women, and the girls making themselves as stiff and ridiculous as possible. All the time six or eight large Chinese gongs were being beaten by the vigorous arms of as many young men, producing such a deafening discord that I was glad to escape to the round house, where I slept very comfortable with half a dozen smoke-dried human skulls suspended over my head.

The river was now so shallow that boats could hardly get along. I therefore preferred walking to the next village, expecting to see something of the country, but was much disappointed, as the path lay almost entirely through dense bamboo thickets. The Dyaks get two crops off the ground in succession; one of rice and the other of sugar-cane, maize, and vegetables. The ground then lies fallow eight or ten years, and becomes covered with bamboos and shrubs, which often completely arch over the path and shut out everything from the view. Three hours' walking brought us to the village of Senankan, where I was again obliged to remain the whole day, which I agreed to do on the promise of the Orang Kaya that his men should next day take me through two other villages across to Senna, at the head of the Sarawak River. I amused myself as I best could till evening, by walking about the high ground near, to get views of the country and bearings of the chief mountains. There was then another public audience, with gifts of rice and eggs, and drinking of rice wine. These Dyaks cultivate a great extent of ground, and supply a good deal of rice to Sarawak. They are rich in gongs, brass trays, wire, silver coins, and other articles in which a Dyak's wealth consists; and their women and children are highly ornamented with bead necklaces, shells, and brass wire.

In the morning I waited some time, but the men that were to accompany me did not make their appearance. On sending to the Orang Kaya I found that both he and another head-man had gone out for the day, and on inquiring the reason was told that they could not persuade any of their men to go with me because the journey was a long and fatiguing one. As I was determined to get on, I told the few men that remained that the chiefs had behaved very badly, and that I should acquaint the Rajah with their conduct, and I wanted to start immediately. Every man present made some excuse, but others were sent for, and by dint of threats and promises, and the exertion of all Bujon's eloquence, we succeeded in getting off after two hours' delay.

For the first few miles our path lay over a country cleared for rice-fields, consisting entirely of small but deep and sharply-cut ridges and valleys, without a yard of level ground. After crossing the Kayan River, a main branch of the Sadong, we got on to the lower slopes of the Seboran Mountain, and the path lay along a sharp and moderately steep ridge, affording an excellent view of the country. Its features were exactly those of the Himalayas in miniature, as they are described by Dr Hooker and other travellers; and looked like a natural model of some parts of those vast mountains on a scale of about a tenth, thousands of feet being here represented by hundreds. I now discovered the source of the beautiful pebbles which had so pleased me in the river-bed. The slaty rocks had ceased, and these mountains seemed to consist of a sand-stone conglomerate, which was in some places a mere mass of pebbles cemented together. I might have known that such small streams could not produce such vast quantities of well-rounded pebbles of the very hardest materials. They had evidently been formed in past ages, by the action of some continental stream or seabeach, before the great island of Borneo had risen from the ocean. The existence of such a system of hills and valleys reproducing in miniature all the features of a great mountain region, has an important bearing on the modern theory, that the form of the ground is mainly due to atmospheric rather than to subterranean action. When we have a number of branching valleys and ravines running in many different directions within a square mile, it seems hardly possible to impute their formation, or even their origination, to rents and fissures produced by earthquakes. On the other hand, the nature of the rock, so easily decomposed and removed by water, and the known action of the abundant tropical rains, are in this case, at least, quite sufficient causes for the production of such valleys. But the resemblance between their forms and outlines, their mode of divergence, and the slopes and ridges that divide them, and those of the grand mountain scenery of the Himalayas, is so remarkable, that we are forcibly led to the conclusion that the forces at work in the two cases have been the same, differing only in the time they have been in action, and the nature of the material they have had to work upon.

About noon we reached the village of Menyerry, beautifully situated on a spur of the mountain about 600 feet above the valley, and affording a delightful view of the mountains of this part of Borneo. I here got a sight of Penrissen Mountain at the head of the Sarawak River, and one of the highest in the district, rising to about 6,000 feet above the sea. To the south the Rowan, and further off the Untowan

Mountains in the Dutch territory, appeared equally lofty. Descending from Menyerry we again crossed the Kayan, which bends round the spur, and ascended to the pass which divides the Sadong and Sarawak valleys, and which is about 2,000 feet high. The descent from this point was very fine. A stream, deep in a rocky gorge, rushed on each side of us, to one of which we gradually descended, passing over many lateral gulleys and along the faces of some precipices by means of native bamboo bridges. Some of these were several hundred feet long and fifty or sixty high, a single smooth bamboo four inches in diameter forming the only pathway, while a slender handrail of the same material was often so shaky that it could only be used as a guide rather than a support.

Late in the afternoon we reached Sodos, situated on a spur between two streams, but so surrounded by fruit trees that little could be seen of the country. The house was spacious, clean, and comfortable, and the people very obliging. Many of the women and children had never seen a white man before, and were very sceptical as to my being the same colour all over, as my face. They begged me to show them my arms and body, and they were so kind and good-tempered that I felt bound to give them some satisfaction, so I turned up my trousers and let them see the colour of my leg, which they examined with great interest.

In the morning early we continued our descent along a fine valley, with mountains rising 2,000 or 3,000 feet in every direction. The little river rapidly increased in size till we reached Senna, when it had become a fine pebbly stream navigable for small canoes. Here again the upheaved slaty rock appeared, with the same dip and direction as in the Sadong River. On inquiring for a boat to take me down the stream, I was told that the Senna Dyaks, although living on the river-banks, never made or used boats. They were mountaineers who had only come down into the valley about twenty years before, and had not yet got into new habits. They are of the same tribe as the people of Menyerry and Sodos. They make good paths and bridges, and cultivate much mountain land, and thus give a more pleasing and civilized aspect to the country than where the people move about only in boats, and confine their cultivation to the banks of the streams.

After some trouble I hired a boat from a Malay trader, and found three Dyaks who had been several times with Malays to Sarawak, and thought they could manage it very well. They turned out very awkward, constantly running aground, striking against rocks, and losing their balance so as almost to upset themselves and the boat; offering a striking contrast to the skill of the Sea Dyaks. At length we came to a really dangerous rapid where boats were often swamped, and my men were afraid to pass it. Some Malays with a boat-load of rice here overtook us, and after safely passing down kindly sent back one of their men to assist me. As it was, my Dyaks lost their balance in the critical part of the passage, and had they been alone would certainly have upset the boat. The river now became exceedingly picturesque, the ground on each side being partially cleared for rice-fields, affording a good view of the country. Numerous little granaries were built high in trees overhanging the river, and having a bamboo bridge sloping up to them from the bank; and here and there bamboo suspension bridges crossed the stream, where overhanging trees favoured their construction.

I slept that night in the village of the Sebungow Dyaks, and the next day reached Sarawak, passing through a most beautiful country, where limestone mountains with their fantastic forms and white precipices shot up on every side, draped and festooned with a luxuriant vegetation. The banks of the Sarawak River are everywhere covered with fruit trees, which supply the Dyaks with a great deal of their food. The Mangosteen, Lansat, Rambutan, Jack, Jambou, and Blimbing, are all abundant: but most abundant and most esteemed is the Durian, a fruit about which very little is known in England, but which both by natives and Europeans in the Malay Archipelago is reckoned superior to all others. The old traveller Linschott, writing in 1599, says: 'It is of such an excellent taste that it surpasses in flavour all the other fruits of the world, according to those who have tasted it.' And Doctor Paludanus adds: 'This fruit is of a hot and humid nature. To those not used to it, it seems at first to smell like rotten onions, but immediately they have tasted it they prefer it to all other food. The natives give it honourable titles, exalt it, and make verses on it.' When brought into a house the smell is often so offensive that some persons can never bear to taste it. This was my own case when I first tried it in Malacca, but in Borneo I found a ripe fruit on the ground, and, eating it out of doors, I at once became a confirmed Durian eater.

The Durian grows on a large and lofty forest tree, somewhat resembling an elm in its general character, but with a more smooth and scaly bark. The fruit is round or slightly oval, about the size of a large cocoanut, of a green colour, and covered all over with short stout spines, the bases of which touch each other, and are consequently somewhat hexagonal, while the points are very strong and sharp. It is so completely armed, that if the stalk is broken off it is a difficult matter to lift one from the ground. The outer rind is so thick and tough, that from whatever height it may fall it is never broken. From the base to the apex five very faint lines may be traced, over which the spines arch a little; these are the sutures of the carpels, and show where the fruit may be divided with a heavy knife and a strong hand. The five cells are satiny white within, and are each filled with an oval mass of cream-coloured pulp, imbedded in which are two or three seeds about the size of chestnuts. This pulp is the eatable part, and its consistence and flavour are indescribable. A rich butter-like custard highly flavoured with almonds gives the best general idea of it, but intermingled with it come wafts of flavour that call to mind cream-cheese, onion-sauce, brown sherry, and other incongruities. Then there is a rich glutinous smoothness in the pulp which nothing else possesses, but which adds to its delicacy. It is neither acid, nor sweet, nor juicy, yet one feels the want of none of these qualities, for it is perfect as it is. It produces no nausea or other bad effect, and the more you eat of it the less you feel inclined to stop. In fact to eat Durians is a new sensation, worth a voyage to the East to experience.

When the fruit is ripe it falls of itself, and the only way to eat Durians in perfection is to get them as they fall; and the smell is then less overpowering. When unripe, it makes a very good vegetable if cooked, and it is also eaten by the Dyaks raw. In a good fruit season large quantities are preserved salted, in jars and bamboos, and kept the year round, when it acquires a most disgusting odour to Europeans, but the Dyaks appreciate it highly as a relish with their rice. There are in the forest two varieties of wild Durians with much smaller fruits, one of them orange-coloured inside; and these are probably the origin of the large and fine Durians,

'Orangs of the Mayas Kassa Race, on a Durian Tree', from Odoardo Beccari, *Wanderings in the Great Forests of Borneo*, Archibald Constable and Co. Ltd., 1904.

which are never found wild. It would not, perhaps, be correct to say that the Durian is the best of all fruits, because it cannot supply the place of the subacid juicy kinds, such as the orange, grape, mango, and mangosteen, whose refreshing and cooling qualities are so wholesome and grateful; but as producing a food of the most exquisite flavour it is unsurpassed. If I had to fix on two only, as representing the perfection of the two classes, I should certainly choose the Durian and the Orange as the king and queen of fruits.

The Durian is, however, sometimes dangerous. When the fruit begins to ripen it falls daily and almost hourly, and accidents not unfrequently happen to persons walking or working under the trees. When a Durian strikes a man in its fall, it produces a dreadful wound, the strong spines tearing open the flesh, while the blow itself is very heavy; but from this very circumstance death rarely ensues, the copious effusion of blood preventing the inflammation which might otherwise take place. A Dyak chief informed me that he had been struck down by a Durian falling on his head, which he thought would certainly have caused his death, yet he recovered in a very short time.

Poets and moralists, judging from our English trees and fruits, have thought the small fruits always grew on lofty trees, so that their fall should be harmless to man, while the large ones trailed on the ground. Two of the largest and heaviest fruits known, however, the Brazil-nut fruit (Bertholletia) and Durian, grow on lofty forest trees, from which they fall as soon as they are ripe, and often wound or kill the native inhabitants. From this we may learn two things: first, not to draw general conclusions from a very partial view of nature; and secondly, that trees and fruits, no less than the varied productions of the animal kingdom, do not appear to be organized with exclusive reference to the use and convenience of man.

During my many journeys in Borneo, and especially during my various residences among the Dyaks, I first came to appreciate the admirable qualities of the Bamboo. In those parts of South America which I had previously visited, these gigantic grasses were comparatively scarce, and where found but little used, their place being taken as to one class of uses by the great variety of Palms, and as to another by calabashes and gourds. Almost all tropical countries produce Bamboos, and wherever they are found in abundance the natives apply them to a variety of uses. Their strength, lightness, smoothness, straightness, roundness, and hollowness, the facility and regularity with which they can be split, their many different sizes, the varying length of their joints, the ease with which they can be cut and with which holes can be made through them, their hardness outside, their freedom from any pronounced taste or smell, their great abundance, and the rapidity of their growth and increase, are all qualities which render them useful for a hundred different purposes, to serve which other materials would require much more labour and preparation. The Bamboo is one of the most wonderful and most beautiful productions of the tropics, and one of nature's most valuable gifts to uncivilized man.

The Dyak houses are all raised on posts, and are often two or three hundred feet long and forty or fifty wide. The floor is always formed of strips split from large Bamboos, so that each may be nearly flat and about three inches wide, and these are firmly tied down with rattan to the joists beneath. When well made, this is a delightful floor to walk upon barefooted, the rounded surfaces of the Bamboo being very smooth and agreeable to the feet, while at the same time affording a firm hold. But, what is more important, they form with a mat over them an excellent bed, the elasticity of the Bamboo and its rounded surface being far superior to a more rigid and a flatter floor. Here we at once find a use for Bamboo which cannot be supplied so well by another material without a vast amount of labour, palms and other substitutes requiring much cutting and smoothing, and not being equally good when finished. When, however, a flat, close floor is required, excellent boards are made by splitting open large Bamboos on one side only, and flattening them out so as to form slabs eighteen inches wide and six feet long, with which some Dyaks floor their houses. These with constant rubbing of the feet and the smoke of years become dark and polished, like walnut or old oak, so that their real material can hardly be recognized. What labour is here saved to a savage whose only tools are an axe and a knife, and who, if he wants boards, must hew them out of the solid trunk of a tree, and must give days and weeks of labour to obtain a surface as smooth and beautiful as the Bamboo thus treated affords him. Again, if a temporary house is wanted, either by the native in his plantation or by the traveller in the forest, nothing is so convenient as the Bamboo, with which a house can be constructed with a quarter of the labour and time than if other materials are used.

As I have already mentioned, the Hill Dyaks in the interior of Sarawak make paths for long distances from village to village and to their cultivated grounds, in the course of which they have to cross many gullies and ravines, and even rivers; or sometimes, to avoid a long circuit, to carry the path along the face of a precipice. In all these cases the bridges they construct are of Bamboos, and so admirably adapted is the material for this purpose, that it seems doubtful whether they ever would have attempted such works if they had not possessed it. The Dyak bridge is simple but well designed. It consists merely of stout Bamboos crossing each other at the roadway like the letter X, and rising a few feet above it. At the crossing they are firmly bound together, and to a large Bamboo which lays upon them and forms the only pathway, with a slender and often very shaky one to serve as a handrail. When a river is to be crossed an overhanging tree is chosen, from which the bridge is partly suspended and partly supported by diagonal struts from the banks, so as to avoid placing posts in the stream itself, which would be liable to be carried away by floods. In carrying a path along the face of a precipice, trees and roots are made use of for suspension; struts arise from suitable notches or crevices in the rocks, and if these are not sufficient, immense Bamboos fifty or sixty feet long are fixed on the banks or on the branch of a tree below. These bridges are traversed daily by men and women carrying heavy loads, so that any insecurity is soon discovered, and, as the materials are close at hand, immediately repaired. When a path goes over very steep ground, and becomes slippery in very wet or very dry weather, the Bamboo is used in another way. Pieces are cut about a yard long, and opposite notches being made at each end, holes are formed through which pegs are driven, and firm and convenient steps are thus formed with the greatest ease and celerity. It is true that much of this will decay in one or two seasons, but it can be so quickly replaced as to make it more economical than using a harder and more durable wood.

One of the most striking uses to which Bamboo is applied by the Dyaks, is to assist them in climbing lofty trees, by driving in pegs in the way I have already described. This method is constantly used in order to obtain wax, which is one of the most valuable products of the country. The honey-bee of Borneo very generally hangs its combs under the branches of the Tappan, a tree which towers above all others in the forest, and whose smooth cylindrical trunk often rises a hundred feet without a branch. The Dyaks climb these lofty trees at night, building up their Bamboo ladder as they go, and bringing down gigantic honeycombs. These furnish them with a delicious feast of honey and young bees, besides the wax, which they sell to traders, and with the proceeds buy the much-coveted brass wire, earrings, and gold-edged hand-kerchiefs with which they love to decorate themselves. In ascending Durian and other fruit trees which branch at from thirty to fifty feet from the ground, I have seen them use the Bamboo pegs only, without the upright Bamboo which renders them so much more secure.

The outer rind of the Bamboo, split and shaved thin, is the strongest material for baskets: hen-coops, bird-cages, and conical fish-traps are very quickly made from a single joint, by splitting off the skin in narrow strips left attached to one end, while rings of the same material or of rattan are twisted in at regular distances. Water is brought to the houses by little aqueducts formed of large Bamboos split in half and sup-ported on crossed sticks of various heights so as to give it a regular fall. Thin, long-jointed Bamboos form the Dyaks' only water-vessels, and a dozen of them stand in the corner of every house. They are clean, light, and easily carried, and are in many ways superior to earthen vessels for the same pur-pose. They also make excellent cooking utensils; vegetables and rice can be boiled in them to perfection, and they are often used when travelling. Salted fruit or fish, sugar, vinegar, and honey are preserved in them instead of in jars or bottles. In a small Bamboo case, prettily carved and ornamented, the Dyak carries his sirih and lime for betel chewing, and his little long-bladed knife has a Bamboo sheath. His favourite pipe is a huge hubble-bubble, which he will construct in a few minutes by inserting a small piece of Bamboo for a bowl obliquely into a large cylinder about six inches from the bottom containing water, through which the smoke passes to a long, slender Bamboo tube. There are many other small matters for which Bamboo is daily used, but enough has now been mentioned to show its value. In other parts of the Archipelago I have myself seen it applied to many new uses, and it is probable that my limited means of observation did not make me acquainted with one-half the ways in which it is serviceable to the Dyaks of Sarawak.

While upon the subject of plants I may here mention a few of the more striking vegetable productions of Borneo. The wonderful Pitcher-plants, forming the genus Nepenthes of botanists, here reach their greatest development. Every mountain-top abounds with them, running along the ground, or climbing over shrubs and stunted trees; their elegant pitchers hanging in every direction. Some of these are long and slender, resembling in form the beautiful Philippine lace-sponge (Euplectella), which has now become so common; others are broad and short. Their colours are green, variously tinted and mottled with red or purple. The finest yet known were obtained on the summit of Kini-balou [Kinabalu], in North-west Borneo. One of the broad sort, Nepenthes rajah, will hold two quarts of water in its pitcher. Another, Nepenthes Edwardsiania, has a narrow pitcher twenty inches long; while the plant itself grows to a length of twenty feet.

Ferns are abundant, but are not so varied as on the vol-canic mountains of Java; and Tree-ferns are neither so plenti-ful nor so large as in that island. They grow, however, quite down to the level of the sea, and are generally slender and graceful plants from eight to fifteen feet high. Without devoting much time to the search I collected fifty species of Ferns in Borneo, and I have no doubt a good botanist would have obtained twice the number. The interesting group of Orchids is very abundant, but, as is generally the case, nine-tenths of the species have small and inconspicuous flowers. Among the exceptions are the fine Cœlogynes, whose large clusters of yellow flowers ornament the gloomiest forests, and that most extraordinary plant, Vanda Lowii, which last is particularly abundant near some hot springs at the foot of the Peninjauh Mountain. It grows on the lower branches of trees, and its strange pendant flower-spikes often hang down so as almost to reach the ground. These are generally six or eight feet long, bearing large and handsome flowers three inches across, and varying in colour from orange to red, with deep purple-red spots. I measured one spike, which reached the extraordinary length of nine feet eight inches, and bore thirty-six flowers, spirally arranged upon a slender thread-like stalk. Specimens grown in our English hot-houses have produced flower-spikes of equal length, and with a much larger number of blossoms.

Flowers were scarce, as is usual in equatorial forests, and it was only at rare intervals that I met with anything striking. A few fine climbers were sometimes seen, especially a handsome crimson and yellow Æschynanthus, and a fine leguminous plant with clusters of large Cassia-like flowers of a rich purple colour. Once I found a number of small Anonaceous trees of the genus Polyalthea, producing a most striking effect in the gloomy forest shades. They were about thirty feet high, and their slender trunks were covered with large star-like crimson flowers, which clustered over them like garlands, and resembled some artificial decoration more than a natural product.

The forests abound with gigantic trees with cylindrical, buttressed, or furrowed stems, while occasionally the traveller comes upon a wonderful fig-tree, whose trunk is itself a forest of stems and aerial roots. Still more rarely are found trees which appear to have begun growing in mid-air, and from the same point send out wide-spreading branches above and a complicated pyramid of roots descending for seventy or eighty feet to the ground below, and so spreading on every side, that one can stand in the very centre with the trunk of the tree immediately overhead. Trees of this character are found all over the Archipelago. I believe that they originate as parasites, from seeds carried by birds and dropped in the fork of some lofty tree. Hence descend aerial roots, clasping and ultimately destroying the supporting tree, which is in time entirely replaced by the humble plant which was at first dependent upon it. Thus we have an actual struggle for life in the vegetable kingdom, not less fatal to the vanquished than the struggles among animals which we can so much more easily observe and understand. The advantage of quicker access to light and warmth and air, which is gained in one way by climbing plants, is here obtained by a forest tree, which has the means of starting in life at an elevation which others can only attain after many years of growth, and then only when the fall of some other tree has made room for them. Thus it is that in the warm and moist and equal climate of the tropics, each available station is seized upon, and becomes the means of developing new forms of life especially adapted to occupy it.

On reaching Sarawak early in December I found there would not be an opportunity of returning to Singapore till the latter end of January. I therefore accepted Sir James Brooke's invitation to spend a week with him and Mr St. John at his cottage on Peninjauh. This is a very steep pyramidal mountain of crystalline basaltic rock, about a thousand feet high, and covered with luxuriant forest. There are three Dyak villages upon it, and on a little platform near the summit is the rude wooden lodge where the English Rajah was accustomed to go for relaxation and cool fresh air. It is only twenty miles up the river, but the road up the mountain is a succession of ladders on the face of precipices, bamboo bridges over gullies and chasms, and slippery paths over rocks and tree-trunks and huge boulders as big as houses. A cool spring under an overhanging rock just below the cottage furnished us with refreshing baths and delicious drinking water, and the Dyaks brought us daily heaped-up baskets of Mangusteens and Lansats, two of the most delicious of the subacid tropical fruits. We returned to Sarawak for Christmas (the second I had spent with Sir James Brooke), when all the Europeans both in the town and from the out-stations enjoyed the hospitality of the Rajah, who possessed in a pre-eminent degree the art of making every one around him comfortable and happy.

A few days afterwards I returned to the mountain with Charles and a Malay boy named Ali, and stayed there three weeks for the purpose of making a collection of land-shells, butterflies and moths, ferns and orchids. On the hill itself ferns were tolerably plentiful, and I made a collection of about forty species. But what occupied me most was the great abundance of moths which on certain occasions I was able to capture. As during the whole of my eight years' wanderings in the East I never found another spot where these insects were at all plentiful, it will be interesting to state the exact conditions under which I here obtained them.

On one side of the cottage there was a verandah, looking down the whole side of the mountain and to its summit on the right, all densely clothed with forest. The boarded sides of the cottage were whitewashed, and the roof of the verandah was low, and also boarded and whitewashed. As soon as it got dark I placed my lamp on a table against the wall, and with pins, insect forceps, net, and collecting boxes by my side, sat down with a book. Sometimes during the whole evening only one solitary moth would visit me, while on other nights they would pour in, in a continual stream, keeping me hard at work catching and pinning till past midnight. They came literally by thousands. These good nights were very few. During the four weeks that I spent altogether on the hill I only had four really good nights, and these were always rainy, and the best of them soaking wet. But wet nights were not always good, for a rainy moonlight night produced next to nothing. All the chief tribes of moths were represented, and the beauty and variety of the species was very great. On good nights I was able to capture from a hundred to two hundred and fifty moths, and these comprised on each occasion from half to two-thirds that number of distinct species. Some of them would settle on the wall, some on the table, while many would fly up to the roof and give me a chase all over the verandah before I could secure them. In order to show the curious connexion between the state of the weather and the degree in which moths were attracted to light, I [made] a list of my captures each night of my stay on the hill.

It thus appears that on twenty-six nights I collected 1,386 moths but that more than 800 of them were collected on four very wet and dark nights. My success here led me to hope that, by similar arrangements, I might in every island be able to obtain abundance of these insects; but, strange to say, during the six succeeding years I was never once able to make any collections at all approaching those at Sarawak. The reason of this I can pretty well understand to be owing to the absence of some one or other essential condition that were here all combined. Sometimes the dry season was the hindrance; more frequently residence in a town or village not close to virgin forest, and surrounded by other houses whose lights were a counter-attraction; still more frequently residence in a dark palm-thatched house, with a lofty roof, in whose recesses every moth was lost the instant it entered. This last was the greatest drawback, and the real reason why I never again was able to make a collection of moths; for I never afterwards lived in a solitary jungle-house with a low-boarded and whitewashed verandah, so constructed as to prevent insects at once escaping into the upper part of the house, quite out of reach. After my long experience, my numerous failures, and my one success, I feel sure that if any party of naturalists ever make a yacht-voyage to explore the Malayan Archipelago, or any other tropical region, making entomology one of their chief pursuits, it would well repay them to carry a small framed verandah, or a verandah-shaped tent of white canvas, to set up in every favourable situation, as a means of making a collection of nocturnal Lepidoptera, and also of obtaining rare specimens of Coleoptera and other insects. I make the suggestion here, because no one would suspect the enormous difference in results that such an apparatus would produce; and because I consider it one of the curiosities of a collector's experience to have found out that some such apparatus is required.

When I returned to Singapore I took with me the Malay lad named Ali, who subsequently accompanied me all over the Archipelago. Charles Allen preferred staying at the Mission-house, and afterwards obtained employment in Sarawak and in Singapore, till he again joined me four years later at Amboyna in the Moluccas.

The Malay Archipelago. The Land of the Orang Utan, and the Bird of Paradise, Macmillian and Co., London, 1869, pp. 49–67.

10
'Men with Tails'

CARL BOCK

Carl Bock, born in 1849, was a Norwegian naturalist and explorer. He had lived and worked in England as a young man, and in 1878, while in London, was sponsored by the Marquis of Tweeddale to make a collecting expedition to Sumatra, in the Netherlands East Indies. Following his Sumatran adventures in 1878–9, he was commissioned by the Dutch Governor-General to organize an expedition to the then little-known interior territories of south-eastern Borneo. This was a semi-official expedition, and Bock was to report on the lands and peoples there and collect specimens.

*In late June 1879 Bock departed from Batavia for
Samarinda, the trading port of the Muslim Sultanate of Kutei.
Between late July and October he made a journey along the
Mahakam River, came back down-river, and then visited
some of the eastern coastal regions northwards to the
Sultanate of Bulungan. On returning to Kutei he left for the
up-river regions again, on 21 November 1879. Travelling up
the Mahakam he then crossed over into the Barito drainage
system and journeyed down to the port of Banjarmasin,
departing from Borneo in March 1880.*

*Although Bock undertook a scientific expedition to Borneo,
his motivations and the tone and style of his narrative are
rather different from those of his compatriot, Carl Lumholtz
(Passage 12), who visited these same regions some three and a
half decades later. Bock was clearly attempting to generate
excitement and a sense of adventure, dwelling on the exotic
and sensational. Of course, one would not wish to underes-
timate the potential dangers of Bock's journey into remote
areas where there was no Dutch administrative presence. But,
unfortunately, Bock was duped by legends of cannibalism
and tailed humans. In his introductory statements he tells us
about 'bloodthirsty Dyaks'. They were 'perfect savages and
inveterate Head-Hunters' (p. 21).*

*The extract which follows comprises Chapter XII and a brief
section of Chapter XIII, describing Bock's adventures in the
Upper Mahakam in November and December 1879. These
passages have been chosen to demonstrate some of the misleading European preoccupations of the nineteenth century
and the tendency, most particularly in popular writings, to
depict the natives of Borneo as existing in a savage state close
to nature. Bock was obviously preoccupied with certain elements of Darwinian evolutionary theory, which for him
entailed the search for a 'missing link'—half human half animal creatures. Yet, despite this extravagance, Bock's account
has some merits. As Reece says: 'It is as a travel book that*
Head-Hunters *must be judged ...' and he concludes that
'Bock was a shrewd observer with a nice sense of humour ...'
(p. x).*[1] *Indeed, some of Bock's narrative is reminiscent of later
travel stories of Borneo, which provide their readers with*

---

[1]R. H. W. Reece, 'Introduction' to Carl Bock, *The Head-Hunters of Borneo*,
Oxford University Press reprint, Singapore, 1985, pp. v–xii.

*adventure, excitement, and exoticism, but which hover un-
easily on the boundary between fact and fiction.*

ON the day after my arrival I was surprised by Kichil
suddenly coming into my room and announcing the
coming of the Sultan [of Kutei], who had informed us
of his intention to stay a few days at Allo. On my asking the
reason for the sudden change of mind, he made an expressive grimace, and said the smell of the fish there was intolerable. He was immediately visited by deputations of the
various *kapalas* and *mantries* [village leaders] in the neighbourhood, who brought with them every afternoon presents
of fruit, eggs, etc. One fine afternoon a dozen of the principal
ladies came to pay *hormat* [respect], and it was rumoured that
his Highness was to select an addition to his hareem. Each
one carried a brass plate or bowl, covered with a white cloth,
the contents of which they seemed to guard jealously from
public gaze, though on inquiry I was informed they contained nothing but rice, eggs, or honey.

Early in the afternoon of the 9th November the Sultan and I
went fishing and shooting up a small tributary of the
Mahakkam called the Djintang, which flows from a lake of
the same name. The stream, like all the rivers in Koetei, goes
in ever-winding ways, and the banks are lined with
magnificent forest vegetation, which prevented our shooting
either birds or animals. It is easy to shoot, but not so easy to
recover the game. The Sultan's hunter wounded a long-nosed
monkey, which fell with a crash from the tree; three or four
of our men hurried up to secure it, but the beast had strength
enough to get away and disappear. When we came to the
lake the Sultan proposed to try our luck at fishing. As the sun
with fiery colours gradually sank below the horizon, the
Sultan directed all the boats to form a circle, and at a given
signal from his Highness, who stood in the bow of his canoe
holding a net in his hands, all the nets were thrown out
simultaneously. This was repeated three or four times, but
without any success. It was too late in the day for fishing, as
the fishes in these lake regions retire before sunset to the
shore, remaining all night amongst the great masses of high
aquatic grass. The Sultan then ordered an immensely long but
narrow net, several hundred feet in length, to be set some
twenty feet from the shore, and a canoe with a crew

remained all night to keep watch, returning the following morning with a rich harvest of fish.

On our return from fishing we found a great number of Longbleh Dyaks of the Modang tribe had arrived to escort us over the frontier, and were moored on the opposite side of the river to that on which our praus lay. Their chief came at once to the Sultan to ask for some rice, which was given to them, and in a few minutes their fires were blazing on the bank. I went across the river the next day to converse with them. The canoes were apparently very old and in want of repair, and the general appearance of the people, and all their accoutrements, indicated a state of poverty compared with that of other tribes. The chief's prau had a square bow, with a painted figure-head of a death's head. The chief himself wore a jacket of leopard-skin, or rather the simple skin of the leopard, with a hole cut in the neck, through which he passed his head, while the head of the skin hung over his chest and the main portion of the skin covered the shoulders and back, the tail almost touching the ground. On the head of the skin, round the edge and inside, were fastened a few conical shells and a large shell of mother-o'-pearl (*Meleagrina*). His head-dress was equally characteristic; it was a conical cap, made from a monkey-skin, with a piece of metal sheathing fastened on the front, and a few rhinoceros feathers stuck in the top. Round his neck he wore several strings of beads. A portable seat of plaited rattan, fastened behind to a tjawat, completed his outfit.

The chief spoke a little Malay, and asked me a number of questions. Why did I want to go this long way to Bandjermasin, when there was a *jallan apie* (a 'fire-road', i.e. a steamship)? What was I going to do with the drawings I made? Why did I ask so many questions about his people and other tribes? I explained to him, by means of an illustrated book I happened to have by me, that I was writing about the natives of Koetei, and so on; and he seemed quite ready to answer my questions, provided I gave satisfactory replies to his, which was not always a very easy task.

His people seemed a fine, well-built, muscular lot of men. Very few of them had any tattoo marks, and the holes in their ears were much smaller than those cultivated by their neighbours of Long Wai. The only ornaments they wore were bead necklaces; but they all carried a number of *tambatongs* [charms] attached to the mandau [sword] girdle. The mandaus were in all cases perfectly plain, without ornamentation of any kind, either on blade, handle, or sheath.

In the cabins of all the canoes were strewn a number of caps and jackets of various materials. The jackets were mostly sleeveless, made of very thick cloth, and padded with cotton wool, as a protection against sword-cuts and poisoned arrows; others were of bear-skin, or monkey-skin (*Nasalis larvatus*), or goat-skin, while a few were made of the bark of a tree, with a little embroidery stitched on.

Detachments from another tribe, the Tandjoeng Dyaks, also came to pay their respects to the Sultan. There were representatives of two branches of this tribe, from Bantang and from Boenjoet. They are not so muscular and tall as the Dyaks in the north of Koetei, but rather slightly built. They do not tattoo as a rule. I only found one with a + on his arm. They make only small holes in their ears, very often wearing no ornaments in them, but they all wear necklaces, mostly a string of beans, called *Boa kalong*. I observed one with a curious necklace composed of red beads and the teeth of a species of bat, set alternately, and producing a pretty effect, the white teeth contrasting well with the vermilion beads. From the necklace, wrapped in a piece of dirty red flannel, hung a talisman. I was very anxious to purchase this necklace. I could not make my wishes understood by signs, or else the man was perverse and would not part with it, and none of my men could understand the dialect, so I took him to the Sultan—who by the way seems to be acquainted with all the languages and dialects spoken in Koetei, and they are numerous enough, except that of the Orang Poonan—and explained what I wanted. His Highness spoke a few words to the Dyak, who agreed to let me have the necklace, but only on the condition that he kept the talisman; he could on no account part with that. He took the necklace off and began to unfold the little piece of flannel, handling his charm as carefully as if it was a precious stone, or at least a bezoar. When he at last hesitatingly held it in his hand for me to see, it proved to be only a tiny piece of yellow wood. I gave in exchange for the necklace three yards of blue cloth.

All these Dyaks wore their hair in a very becoming fashion, reminding me much of some of the so-called chignons in fashion among the ladies some years ago in Europe. The hair is cut short below the occiput, while on the crown it is allowed to grow to a great length, sometimes reaching to the knees. This long hair is rolled up in chignon fashion and fastened by a sort of head-covering made of bark, resembling the New Zealand tappa. This is the only instance in which I saw this material dyed—coloured red, blue, or yellow. The *tjantjoet* or *tjawat* is mostly of the same material, being preferred to cloth, as being more durable.

I was struck by the fact that none of the mandaus worn by these men, again, were ornamented, all being perfectly plain. They often begged for tobacco, and one day I let them taste a drop of brandy, which they did not like. I was told that among the Tandjoeng Dyaks there are only a couple of houses in each village, but so large as to contain between them the whole population of 400 to 500.

At Moeara Pahou I had expected to meet a party of the much-dreaded Tring Dyaks, a branch of the Bahou tribe, having sent a messenger from Kotta Bangoen with a request that the Rajah would meet me here with a number of men and women of his tribe. I had taken the precaution to send a present of a picol of rice and some fish in proof of my friendly intentions, and promised that, if the chief would allow me to make a few sketches of his people, he and they should be liberally rewarded. It was four days' journey from Kotta Bangoen to the Tring settlement, and I was not surprised to find on reaching Moeara Pahou that none of them had arrived. But when two or three days had elapsed and still no Trings appeared on the scene, I determined to go myself to their kampong; but the Sultan and all the people said it would not be safe to do so: the people were cannibals, and were hated as well as feared by all their neighbours, and they might possibly think that the large force which the Sultan had collected here was brought together for the purpose of attacking them, especially as some of the assembled tribes were unfriendly to the Trings. I explained that I must see them, having heard so much about their atrocities and cannibalism. The Government would expect me to report upon these savages; and I should be to blame if I did not see them, both men and women. So the Sultan sent a canoe, with a reliable man in charge, to request the Trings to put in an appearance. Four, five, six days passed, and still no Trings came, and, more strange, no canoe returned. Were the crew killed and eaten? The Sultan could not sit still under such a possibility, and sent another large canoe, well armed, and in command of a Kapitan, who came back in three days, bringing the first envoy with him, and some forty Trings besides, including four women.

The men seemed to exhibit in their bearing a strange mixture of shyness and suspicion. They wore a tjawat, or waistcloth, of bark, and a head-covering of the same material. They were slightly tattooed—a small scroll on the arm or calf of the leg; and they all had their ears pierced, and the holes enlarged, though only a few of them wore any ornament, generally a wooden cylinder, in the ears. I sketched one of the men in war-costume, bribing him with a couple of dollars to go through the war-dance. Running round and round, stamping his feet heavily on the ground, shouting at the top of his voice, flourishing his mandau as if striking an imaginary foe, and then guarding himself with his shield, he gradually became so excited and furious in his movements, cheered on by the cries of his companions, that I was not sorry to think that I was not witnessing a *pas de deux*.

The women were much more elaborately tattooed than the men, the whole of the thighs and the hands and feet being covered with blue patterns. Their dress consisted of a sort of petticoat, either of a blue or neutral tint, fastened round the hips and reaching to the ankles, bordered at each end with a red piece of cloth. Many of them wore round the waist several strings of large beads of a turquoise, dark blue, or yellow colour. These I found were highly prized, being very old (from *tempo doelo*), and no longer procurable. Round their necks was also a profusion of beads. Of head-coverings they wore two sorts; the one a conical hat, without a crown, covered with fine bead-work, and bound round the edges with a strip of red flannel; the other, merely a narrow band of red flannel, beaded at intervals in regular patterns, and fastened with a button behind.

The lobes of the ears were pierced, sometimes in no less than three places in addition to the large central slit, the principal holes being enormously enlarged by the weighty tin rings hanging in them. The kapitan told me that these people live in large houses several hundred feet long, but extremely dirty inside, and of a wretched appearance outside. The houses, he said, were literally full of skulls taken by the tribes in their head-hunting expeditions. I noticed that the other Dyak tribes did not go near the Trings during their stay at Moeara Pahou, not disguising their fear of them, and their disgust at their cannibal practices.

These people speak quite a distinct language, and none of the Tangaroeng Malays could understand a word that they said; fortunately the old Boegis kapitan who brought them could converse fluently with them, having lived some years amongst them as a sort of tax-officer for the Sultan, and with his assistance I was enabled to obtain much information. Among the visitors was an old priestess, who gave full details concerning the religious beliefs, etc. of the tribe. This information was elicited by the kapitan, and interpreted by him to a Malay writer, who took down the statements on the spot. These statements have since been translated for me, and are embodied in the chapter on the religious rites of the Dyaks.

· This priestess allowed me to take her portrait. The most striking feature is the enormous length of the loops formed in the lobes of the ears, from which heavy tin rings were suspended. She allowed me to accurately measure this monstrous deformity.

Next, the absence of eyebrows will be noticed. The eyebrows are either entirely wanting or very scanty in all the members of the tribe, who pull them out, considering their absence a mark of beauty.

The elaborate tattooing on the thighs is also a striking feature. The shortness of the hair, again, is in contrast to the length to which the women of all the other tribes allow their hair to grow; and the colour of the skin is slightly lighter than that prevailing among the Dyaks, the Orang Poonan alone excepted.

This priestess in the course of conversation told me— holding out her hand—that the palms are considered the best eating. Then she pointed to the knee, and again to the forehead, using the Malay word *bai, bai* (good, good), each time, to indicate that the brains, and the flesh on the knees of a human being, are also considered delicacies by the members of her tribe.

Having interviewed this priestess, I had the honour of an introduction to the famous, or infamous, chief of the cannibal Dyaks, Sibau Mobang. He came into my house one day, accompanied by his suite of two women and three men, and

I hardly know whether host or visitor felt the more uncomfortable. His personal appearance bore out the idea I had formed of him by the reports I had heard of his ferocity and the depravity of his nature; but I was hardly prepared to see such an utter incarnation of all that is most repulsive and horrible in the human form.

As he entered my floating habitation he assumed a sort of air of hesitation, almost amounting to trembling fear, which added to, rather than detracted from, the feelings of repulsion with which I viewed him. He stood for a moment or two, neither moving nor speaking, watched me narrowly when I pretended not to be looking at him, and then sat down quietly a couple of yards from my feet. He is a man apparently about fifty years of age, of yellowish-brown colour, and a rather sickly complexion. His eyes have a wild animal expression, and around them are dark lines, like shadows of crime. He is continually blinking his eyes, never letting them meet those of his interlocutor, as if his conscience did not allow him to look any one straight in the face. His face is perfectly emaciated, every feature shrunken and distorted. The absence of teeth in the gums gives the bones an extra prominence. A few stiff black hairs for a moustache, and a few straggling ones on his chin, add to the weird look; his ears hang down low, pierced with large holes two inches in length. His right arm, on which he wears a tin bracelet, is paralyzed, and he is unable to open the right hand without the assistance of his left, lifting each finger separately, and closing them again with little less difficulty. For this reason he wears his mandau on his right side, and the many victims that have fallen to this bloodthirsty wretch during the last few years he has decapitated left-handed. At that very time, as he sat conversing with me through my interpreter, and I sketched his portrait, he had fresh upon his head the blood of no less than seventy victims, men, women and children, whom he and his followers had just slaughtered, and whose hands and brains he had eaten.

He told me his people did not eat human meat every day—that was a feast reserved for head-hunting expeditions; at other times their food consisted of the flesh of various animals and birds, rice, and wild fruits. For a whole year, however, they had had no rice, owing to the failure of the crops. When I heard this I told Kichil to bring forward a large kettle of rice which was boiling, and to place it before my guests, together with some salt. The eagerness with which they ate the rice, rolling it first between their hands so as to form solid rolls, bore out the statement that they had lately been kept on very 'short commons' indeed.

The whole time he sat in my room Sibau Mobang seemed very grave, and kept incessantly turning his head away from me, so that it was not difficult to get a portrait of him in profile. His grim visage, his still more grim manner, made me wonder whether he could ever laugh. The idea seemed horribly ludicrous; I tried however to get a a smile on his countenance, but without success, until, when I had finished my sketch, I handed it to him to examine. He scrutinized it closely, then looked at me for the first time full in the face, and actually smiled, a ghastly grim smile, horribly suggestive of nightmare. He made signs that he wished to keep the sketch, but I made him understand that I could not let him have it. I gave him, however, various presents, and two dollars to each member of the party whom I had had the privilege of sketching, besides a picol of rice, some strings of beads, and twenty-four yards of calico to divide between them. Sibau Mobang gave me in return two human crania, trophies of his head-hunting excursions—one that of a male, the other of a female, but both, as usual, wanting the lower jaw; they were wrapped up in pisang [banana] leaves. He also with some reluctance gave me a *kliau* (shield), of the ordinary soft wood, painted in grotesque patterns, and ornamented with tufts of human hair most ingeniously stuck on. Such a shield is considered a great treasure, being decorated with hair taken from human victims.

This cannibal, however, is not the chief Rajah of the Tring Dyaks. Their nominal ruler is Raden Mas, a chief who, at the instance of the Sultan, his suzerain, gave up cannibalism in order to embrace the Mohammedan religion, and enjoy the advantage of a plurality of wives. He is very rich, very powerful, and very independent of the Sultan, who is obliged to humour him very much to keep him on good terms, and who, on the occasion of his supposed conversion to Islam, gave him the title he now holds of Raden Mas: *Raden* = noble; *Mas* = gold. The latter term refers to the stores of gold-dust which he is reputed to possess, hidden away in his village.

The Raden was invited to join the Sultan's suite and accompany him to Bandjermasin. He has large eyes and prominent cheekbones, and the unsightly long holes in the lobes of the ears; but his general appearance is by no means repulsive. Dressed in a neat cotton or silk jacket, with gold buttons, and a pair of short Boegis trousers, he was, compared with his second in command, Sibau Mobang, quite a gentleman.

*   *   *

On my return to my prau I found the Sultan, with his numerous fleet, had followed me. The Dyak praus were all moored on the opposite side of the river, keeping apart from the Malays. The Sultan's large prau, flying the yellow standard with the tiger *rampant*, was fastened close to mine; and in the evening we dined together on a raft in the river, on which a table and two seats were hurriedly constructed. We were now on the confines of the territory of the Sultan, who was anxious to make as large a display of force as possible. He had recently, he told me, lodged a complaint to the Dutch Government against the Doesoen tribes, whose territory we should now enter, and who were the great rivals of the Long Puti Dyaks, making frequent incursions into their territory.

More interesting, perhaps, was his statement that we were now within a short distance of the country in which the tailed race of men lived. The existence of these people was the common talk, not only here, but all the way down to Tangaroeng, and they were variously stated to dwell in Passir, and on the Teweh river. We 'discussed' these people over a basket of durian. This was the first time I had tasted this celebrated fruit. The smell of the fruit was not very appetizing, and the flavour—to my taste—resembled that of bad onions mixed with cream.

The Sultan's men were short of provisions, but their commissariat was replenished by a heavy requisition which the Sultan made on the people of this the largest settlement at the furthest extremity of his dominions. So there was great feasting, not only on board the praus, but on shore, where the people killed goats and buffaloes, and distributed a slice to every person in the village. In the evening the hill was ablaze with innumerable fires, round which gambling and card-playing, smoking and sirih-chewing, alternated with the operation of cooking and eating the meal; while the dogs kept up a continual barking and fighting over the bones that were thrown to them as the feast proceeded.

The conversation about the tailed race brought back to my mind various rumours of the existence of this 'missing link' in the Darwinian chain which had reached me at different times during my travels in Borneo, and I determined if possible to settle the point one way or the other. The question has often occurred to me whether Mr Darwin received the first suggestion of his theory of man's simian descent from the fables concerning the existence of tailed men which obtain credence among so many uncivilized people: or whether the natives of the Malaya Archipelago and the South Sea Islands, having read the 'Descent of Man', have conspired together to hoax the white man with well-concocted stories of people possessed of tails, living in inaccessible districts, and maintaining but slight intercourse with the outer world. It is certainly a curious fact that similar stories exist, not only in Borneo, but in other islands in the South Pacific—New Britain for instance, where missionaries have more than once been tempted into hazardous expeditions in search of the great physiological prize, the missing link in the chain of evidence proving the descent of man from monkey.

I made inquiries in the village, and found a strong general belief in the existence of people with tails in a country only a few days' journey from Long Puti. Such definite statements were made to me on the subject that I could hardly resist the temptation to penetrate myself into the stronghold of my ancestral representatives. Tjiropon, an old and faithful servant of the Sultan, assured me, in the presence of his Highness and of several Pangerans, that he had himself some years ago seen the people in Passir. He called them 'Orang-boentoet'—literally, tail-people. The chief of the tribe, he said, presented a very remarkable appearance, having white hair and white eyes—a description which exactly agreed with one I had received some time previously from a young Boegis, when travelling by steamer to Samarinda from Pare Pare in Celebes. As to the all-important item of the tails, Tjiropon declared with a grave face that the caudal appendage of these people was from two or four inches long; and that in their homes they had little holes cut or dug in the floor on purpose to receive the tail, so that they might sit down in comfort. This ludicrous anti-climax to the narrative of the trusty Tjiropon almost induced me to discredit the whole story. At any rate, I thought, the Orang-boentoet must be in a very high state of development—or rather, perhaps, in the last stages of retrogression—if the extremely sensitive prehensile tail of the spider-monkey has so lost its elasticity in these people as to incommode its wearer to such a degree. The Sultan, however, was highly impressed with the truth of Tjiropon's story. He had often heard that there were among his neighbours, if not even among his own subjects, a tribe with tails; but he had hitherto discredited the rumours. 'Now', he said, 'I do believe there are such people, because Tjiropon has told us. I have known him for twenty years, and he dare not tell a lie in my face, in presence of us all.'

So we asked Tjiropon if he would go and pay another visit to his former friends, and bring one or two of them to introduce to us. He was at first unwilling to go, on account of the disorder existing in Passir, and of a predilection which the inhabitants were alleged to have for poisoning strangers. But a present of 600 florins and of a suit of clothes, and the promise of a reward of 500 florins if he brought a pair—or couple, should I rather say—of tailed people safely to Dutch territory, overcame his scruples. The Sultan decorated the clothes I gave him with a set of silver buttons, adorned with his coat of arms, so that he might present a respectable appearance before the Sultan of Passir, to whom he was furnished with letters of introduction. Thus armed with authority, and with an escort of fifteen men, Tjiropon set out on his expedition, with orders to *rendezvous* at Bandjermasin.

Having despatched Tjiropon on his important mission, we continued our journey up the stream, leaving Long Puti at six a.m. on the 20th December; but we had not gone far before the river became very narrow and shallow, and the current so strong that we proceeded with difficulty. Having stemmed two rapids, we were confronted by a series of falls, which necessitated our unloading the canoes, and carrying the luggage about a mile through the forest, the river meanwhile taking a long sweep of two or three miles to the right. Unfortunately, I was myself added to the list of *impedimenta*, being seized with a sudden attack of fever, and had to be carried in my hammock through the forest by four Dyaks. In the meantime our canoes had been safely hauled over the falls, and through the rapids above, and were in readiness to take us on to Moeara Anan, which we reached in the evening.

The Head-Hunters of Borneo: a Narrative of Travel up the Mahakam and down the Barito; also Journeyings in Sumatra, Sampson, Low, Marston, Searle and Rivington, London, 1881, pp.128–36, 143–5.

11
An Early Tourist in Sarawak

FREDERICK BOYLE

Frederick Boyle, a Victorian gentleman and Fellow of the Royal Geographical Society, departed from England on 1 February 1863 in the company of his brother. They left Suez in early April and took a month to reach Singapore, where they were delayed for a further two months. Far from being an explorer or scientific observer, Boyle, in many respects, was a latter-day tourist. He wanted to satisfy his curiosity and amuse himself. Sadly, Singapore neither satisfied nor amused him, and a few of his comments give us some of the flavour of the book. Singapore, Boyle tells us, was 'Decidedly ... the least sociable colony of England' where 'No public amusement whatever exists' (p. 2).

From Singapore, Boyle eventually reached Kuching in Sarawak, on Sir James Brooke's mail steamer, the Rainbow. *He then travelled quite extensively in Sarawak, taking in, as a tourist would, the main sights: the Chinese gold mines, the Land Dayak tribes reached by a 'sampan' ride up the Sarawak River, the coastal peoples of Bintulu and Mukah, and the notorious Sea Dayak 'pirates' and head-hunters of the Saribas and Skrang Rivers.*

It is instructive to compare Boyle's narrative with the serious observations of such explorers as Carl Lumholtz. Although often informative, and certainly eminently readable, Boyle's European prejudices and paternalism show through. Here are just two examples: 'Towards afternoon we reimbarked in our

*sampan, after distributing our present of beads and tobacco,
and sketching a few of the natives' (p. 44), and 'Most of the
Dyak tribes have no marriage ceremony at all, unless the
simultaneous drunkenness of every male in the house may so
be called' (p. 212).*

*The following extract comprises Chapter IV, in which Boyle
describes his journey to the Rejang River in the gun-boat, the*
Jolly Bachelor, *in the company of officers of Sir James Brooke.
He also provides a brief description of the Kanowit Dayaks of
the Lower Rejang basin, whose womenfolk Boyle found to be
'charming beings'.*

THE *Jolly Bachelor* is the smaller of Sir James Brooke's
gun-boats, and is commanded by Capt. Micheson. Her
burthen is about forty tons, and her origin is lost in the
distance of an antique age. Sarawak deprived of the *Jolly*
would be like America without the memory of Washington—
her history would die with its founder. In former times the
Jolly was celebrated for the incredible number of scorpions
and centipedes which infested her timbers; since her last re-
surrection these plagues have disappeared, and the *Jolly* has
no longer a claim to personal glory, save in the events which
she has seen and participated.

She is not a fast boat, nor can she be called comfortable;
the voyager is never quite satisfied, as he reflects upon her
hard service and astounding longevity, that she may not
come to a sudden end like the 'deacon's chaise', and cast him
forth upon the deep. Nor has she that steadiness and sobriety
of movement which should accompany virtuous old age, but
certainly she has some of its other comforts—love, and honour, and troops of friends. No one has been to Sarawak but
knows the *Jolly*, and can remember many a pleasant hour
passed in her low cabin, as she lay becalmed in some silent
creek or river, where the thick jungle was all around, the
water rotting among the mangroves beneath, and the deep
blue sky overhead.

But running up the N. W. coast at the change of the mon-
soons was not quite so pleasant as the dreamy river travel.
The weather was very bad, as was to be expected at the season, and we had several accidents. One night, as we dashed
up the Rejang river in a terrible squall of wind and rain, I
stood in the companion looking for a break in the shroud-like

sky. Suddenly I heard a wild cry overhead, and at the same moment a dark mass shot down before my face, and dashed with fearful violence against the side. We hastened to raise the horrid-looking heap from the sloppy deck, and found it to be the body of an unfortunate sailor who had fallen from aloft. His thigh was broken, and his head badly cut and bruised. In the morning we discovered that his skull had drilled a round hole in the bottom of the sampan which was hauled up alongside. Had it not been for this, however, the poor fellow must have gone overboard, and the tide was running out very fast. Even if he had escaped the sharks, the chances were much against his making shore in such a night and such a sea.

When daylight came, we endeavoured to set the broken thigh; but the anatomical difficulties which we found seemed insurmountable, and we satisfied ourselves with binding it up in bamboo splints, and putting the man ashore at a village where he said he had friends.

A few days after this event—passed in an equal proportion of sharp squalls and merry calms, blazing days and soft starlit nights, when we used to sit out on deck singing nigger songs, and trying to talk sentiment about the Southern Cross, which glimmered down on the horizon—we reached Kennowit [Kanowit]. This town, situated about 150 miles up the Rejang river, is inhabited principally by a tribe of the same name, formerly very powerful in this part of the island, but now rapidly diminishing in numbers and importance. The Kennowits, indeed, complain that they are in danger of extinction from the Dyak [Iban] clans surrounding them, and that the latter leave them no ground on which to plant their rice. They have never been well-disposed towards the Rajah's government, and engaged heartily in the conspiracy of 1859 against it, signalising themselves therein by the murder of their Residents, Messrs Fox and Steele, as these gentlemen were strolling unarmed outside the fort.

In appearance the Kennowits contrast badly even with the Land Dyaks, and are far inferior to the sea tribes. I have already observed that they belong to one of the five aboriginal races, but in language and characteristics they are so closely allied to the Kyans as almost to be merged in that powerful tribe.

The Kennowits have the Tartar cast of face and figure

which characterises more or less every race in the island, and is especially noticeable among the Malays. Their colour is yellow, their eyeballs small and prominent, the ridge of the nose between the eyes is almost imperceptible, the mouth is large and shapeless, the cheek-bones project enormously, the nostrils are wide and flat, and the expression generally sullen and malignant. The above description may be held to refer also to the Kyans, with whom the Kennowits are closely connected; but the Dyaks, though perhaps of the same original stock, are much more agreeable in feature and expression. But it is principally in point of language that the five races show such marked dissimilarity.

The Kennowits, and their kindred the Kyans, tatoo the chest in pale blue lines with an occasional streak of scarlet. Many of the arabesques are very intricate and beautiful, but I never saw them attempt to delineate the figure of any animal. Sometimes they give themselves beard and whiskers in blue tracery; naturally their faces are quite smooth, as is the case with all the natives of the far East. Both Malays and Dyaks consider tatooing to be a sign of cowardice, for, say they, a brave man requires no adventitious aid to make him terrible, and in fact the Kennowits are not highly esteemed for courage.

Another of their customs is the enlargement of the lobe of the ear to an enormous size: I have seen cases where an orange could be passed through the orifice with the greatest ease.

The costume of the men is precisely the same as that I have described as prevailing among the Land Dyaks, and their habits are similar. The women wear the bedang or short petticoat, but when working in the sun they put on a quilted jacket, open down the front. On great occasions, such as a deputation to their Resident, they come out in full Malay costume.

The Rejang river upon which the town of Kennowit stands, is the largest stream in the island; it is more than half a mile wide opposite the town, which is situated about three hundred miles from the sea. The fort stands on a hill overlooking the water, and, like all the other out-stations of Sarawak, is a wooden building of two stories. The lower part has no windows, and the only entrance is by a ladder to the first floor. It is well provided with cannon, and there is a plentiful supply of firearms upon the racks; mostly, however, smooth bores and very ancient.

The morning after our arrival, as I looked through the lattice-work under the eaves of the fort, I beheld a most astonishing phenomenon. Immediately below me was a

'Tatooed Kenowit, with Pendulous Ear-lobes', *Illustrated London News*, 10 November 1849.

round circle of straw nearly five feet in diameter—beyond that another, and a dozen more beyond that, in due diminution of perspective. The objects were in motion winding along the path in a horribly business-like manner. I remained speechless for a while, but when the first had passed on some little distance, beneath the extraordinary straw covering I perceived the body and long black hair of a woman. Another and another was disclosed, and the mystery was solved; the wives of the Kennowits were coming to pay their respects to Mr Cruickshank, their Resident, on his safe return. The wonderful structures which so overwhelmed me when seen from above, were neither more nor less than their hats, and indeed they could not well have been more, though easily less.

Most of the women were of astonishing age and amusing ugliness. One little girl, however, who was certainly not more than four feet high, though full-grown and married, seemed rather good-looking.

These charming beings squatted down in a semi-circle in Mr Cruickshank's bedroom, and instantly began to express their sentiments with a freedom and decision which told well for their household independence. They showed, or at least professed, their contempt for the presents offered them without the slightest hesitation. The deputation was dressed in long Malay jackets of blue cotton, very tight, very shiny, and provided with a considerable number of gold or gilt buttons. For this festive occasion only they wore Malay sarongs about their waists, falling down to the ankles; their whole appearance indeed far more resembled that of Malays than of Dyaks.

The worthy ladies were treated with scant ceremony according to European ideas. Mr Cruickshank went in and out of the room all the time, had a game at play with his big monkey, and finally left the women entirely alone while he went on board the *Venus*, which was lying in the middle of the river. But the deputation thought nothing of it apparently; they squatted on their haunches in perfect contentment whether their Resident was present or not, and harangued us with great cheerfulness when he had finally disappeared. We could not understand a word they said, nor for that matter could anybody else except Mr Cruickshank, who has necessarily studied their language.

In the afternoon our party was invited by a famous Kennowit warrior named Joke—a faithful friend to the Rajah's government—to be present at a great festivity to be held at the house of the Orang Kaya, in honour of the Tuan Mudah's arrival. After dinner accordingly we embarked upon the moonlit river in a sampan or native boat. But the sampan was very small, and our party rather large, from which two causes it resulted that, long before we reached our destination, everyone became uneasy. We first put all the heavy men in the middle, then we divided them: then we devised another mode of balancing the craft, in the midst of which she began to sink and we made a desperate rush for the bank, to which we were fortunately close. The sampan went down just as the last of us jumped out. I may mention, in passing, that the Rejang is celebrated, even among the rivers of Borneo, for the peculiar malevolence of its sharks and alligators.

When we landed the moon had disappeared behind a bank of clouds, and the darkness was almost visible. The houses were at some distance from the spot where we were; there was no path, or in their excitement our guides had missed it, and we were obliged to make our way through the underwood as best we could. For my own part I will here make a confession. The reason that I reached the Orang Kaya's house in such coolness and comfort was that I performed the journey upon the back of a stalwart little Malay fortman, who whispered the proposition into my ear as we landed. I took care, however, to preserve my dignity by alighting just before reaching the door.

We found the Orang Kaya's house to be raised quite twenty feet from the ground, though it was not the tallest in the town. To gain this elevation the visitor is obliged to climb up a notched log provided with a slight handrail to assist the balance. When we had mounted, we found ourselves in a sort of verandah, upon which opened the door of a large irregular chamber crammed with natives. The ceremony of hand-shaking is thought much of by them, and every warrior was anxious to go through it with us. When they were all satisfied, we were conducted to one end of the apartment, and there sat down with our backs to the wall. In case of accidents, our party was very strong, consisting of six Europeans well armed, and twelve Malays in the Rajah's uniform, tight blue jacket, red sash, and white trousers. And this enumeration does not include 'Din, the Tuan Mudah's pretty little Malay boy, whom I saw in a prominent position, girt with a sabre so long that he must have drawn it over his shoulder after the manner of the crusaders.

After waiting some time in considerable impatience, a board was laid down in front of us, upon which dancing was to take place. The next moment a warrior leapt upon it with that correct spring we have seen so often on a more luxurious stage. The music opened up, the performer began to dance, and the scene became striking enough.

The immense apartment was full of queer corners and recesses, so crammed with spectators as barely to leave a space clear for the dancing. Such light as there was came from a number of tall bronze braziers, of design and workmanship by no means contemptible, which were filled with some odoriferous wood. Being principally disposed near ourselves, and emitting a dense smoke and a dazzling red flame, they enabled us only to catch an occasional glimpse of a multitude of dun figures squatted on the ground, whose bright eyes gleamed with excitement in the obscurity. Round the semi-circular space behind us and on our flanks were ranged cloths and sarongs of brilliant colour, while on the board in the centre danced the warrior in naked dignity. On our right were the Malays, whose savage features and tasteful costumes seemed to make the background still more striking and barbarous.

So far as we could ascertain, there was no particular significance in the first dance. It was very slow and solemn, with much swaying from side to side, and stooping, and turning about, all, however, in excellent time to the music. After a while another warrior joined the first, and both crouched down on their haunches in the usual Dyak fighting position, and went through figures apparently as much *de rigueur* as those of a European quadrille. Both were completely armed with sword and shield, and the strength and agility displayed were rather striking, but otherwise the performance appeared to 'drag'.

When these two had perspired till they seemed likely to melt away entirely, our friend Joke, who had been looking on with manifest impatience, suddenly sprang up and took his place on the board.

Joke was a little man,—probably not over five feet two

inches in height,—but with a breadth of shoulder and depth of chest which would not have disgraced a life-guardsman. There was not an ounce of flesh to spare on his yellow-brown body, which glistened with health and condition, and his broad chest was covered all over with graceful blue arabesques. The huge lobes of his ears hung on his shoulders, and through each two broad brass rings were passed, three inches in diameter and half an inch thick. He prepared for his part by putting on a war-cloak of wild bull's hide, adorned with feathers of the rhinoceros hornbill, and trimmed with red cloth and panther-skin. More feathers of the same bird fluttered about his war cap of monkey-fur and fell into his little prominent eyes, which glistened with gin and excitement. He was about to show us a dance, designed to represent the principal events in the life of an orang-outang or mias, and it must be evident to all thinking individuals that sword, and shield, and war cap are essentially requisite for the correct delineation of that animal's habits.

The contortions through which Joke put his person and his features are beyond description; suffice it to say that, by the aid of much nature and a little art, he managed to give himself something of the personal appearance of an orang-outang. But if his imitation of the animal's postures and general habits were at all true to nature, I can only say that in Europe we have much to learn on this subject. The performance was brought to rather an abrupt conclusion, and the mias was left in a peculiarly uncomfortable posture, with its cap over its eyes. Then Joke volunteered to give us a deer-dance in like manner. This also was entirely opposed to the ideas usually current as to this animal's habits, and might have led an inexperienced traveller to conclude that a deer in Borneo was engaged through its whole existence in beating time to music on its hind legs, with a drawn sword in its fore-paw.

Joke deserted his deer in a very sudden manner, just as the animal was apparently about to volunteer a song. A glass of neat gin restored him to his pristine vigour, and, in company with another chief, he gave us one of the traditional dances of the Kennowits. Arranged in all the bravery which delights the military mind in most quarters of the globe, these two chiefs made a great fight in the heroic style,—musical time, stamping of feet, and turning of backs. The slow activity, if I may so express myself, which they showed in leaping from side to side in the constrained position in which Dyaks fight, seemed very remarkable. After a while Joke, who always contrived to be the hero of the scene, finding himself evenly matched, commenced, in pantomime, to make spikes under cover of his shield, which was adorned with figures in blue and red paint, and with streaming locks of human hair. The spikes consisted of bamboo stakes, sharpened and thrown about the ground; a most dangerous weapon against a bare-footed enemy, for the green bamboo is as hard as a steel blade, as I know by painful experience. In the end Joke was successful, and his foe, lame and helpless, was dispatched after a courageous *pas seul* upon one leg.

Then ensued a wild fandango of triumph round the body, while the drums and gongs composing our orchestra beat like mad things. Presently, amid the yells of the audience, Joke tore off his enemy's head—that is, his cap—and danced about with it. On more attentive examination, the miserable man recognised the features of his brother, and howled. After simulating grief and horror, much as it is done at the Italian Opera, he adopted a bold resolution, seized his brother's shoulders, spat furiously into the cap, and thrust it upon the dead man's head. Upon which the brother leapt to his feet, and the two executed a 'pas de congratulation'. This was the hit of the evening, and the yells were awful.

After resting awhile, and drinking much more raw spirit than was good for him, our indefatigable friend leapt again upon the stage, like a giant renovated with gin.

This time he gave us a piece of pantomime representing a Kennowit jungle campaign. It was admirable both in design and execution, comprising the various incidents of an un-successful expedition in search of heads, such as ten years ago occupied half the existence of these fellows;—the other half being passed in successful excursions. It behoved the performer to do his best, for the audience were critical upon the representation of events which, passing away from the domain of the 'chic' and the romantic, came within the experience of every one of them, and Joke, who was a brave and renowned warrior, simply drew upon his memory for the 'points' of the performance.

He began with a spirited strut along his board, representing a proud departure from home with his companions. Then he crept stealthily along in an enemy's country. The surprise followed, a general fight, and the death of all his comrades. Then he wandered about, lost in a hostile jungle, afraid to light a fire, and starving. Finally, with a last faint stamp in time to the music, he lay down to die, but rousing himself for a final effort, crawled home and celebrated his escape with a lively jig. The whole pantomime was exceedingly clever, and required scarcely a word of explanation. Joke is celebrated for this dance, and for the one with his brother which I have described, and when taken to Kuching by Mr Cruickshank, he is much supplicated for a performance, but it is generally to no purpose.

The dancing, with the intervals between, had now lasted about three hours, and the atmosphere of the hut was curious. Also every one of the warriors was drunk, or in a fair way to become so, and the dignity of the Government officers began to warn them to move. Three aged and hideous hags, arrayed in long blue coats of silk with Malay 'sarongs', and gold ornaments, attempted to get up a female exhibition, accompanied by much waving of arms and shaking of stomachs, but somehow they could not at all agree about the right mode of commencing, and we would not wait for the end of the very voluble argument in which they instantly engaged.

So about midnight we sallied out from the smoky gin-sodden atmosphere of the hut, and returned in safety on board the *Venus*. As we sat on deck smoking a last cheroot and watching the ripple of the moonlight on the swift river, the yells of the Kennowits came faintly to our ears, showing that the excitement of the festival had by no means abated after our departure.

Adventures among the Dyaks of Borneo, Hurst and Blackett Publishers, London, 1865, pp. 73–87.

Travellers during
the Late Colonial Period

12
Scientific Exploration in the Upper Mahakam Area of South-eastern Borneo

CARL LUMHOLTZ

Carl Lumholtz was the quintessential professional explorer.[1] Born in Norway in 1851, he spent much of his later life in America. He was educated at the University of Christiania (Oslo), taking a first degree in theology, and then studying zoology. Between 1880 and 1884 he lived and travelled among aborigines in Queensland, Australia, collecting specimens for the zoological museum at the University of Christiania. It was during his time in Australia that Lumholtz became interested in the study of culture, and he turned increasingly to the academic discipline of anthropology.

Lumholtz travelled to America in 1890, and during the next eight years undertook four major scientific expeditions to the desert areas of Mexico. He organized a further two expeditions there in 1905 and 1909–10. Lumholtz then turned his attention to New Guinea, intending to explore remote parts of the interior. He first went to Borneo in 1914 to secure bearers and boatmen for his New Guinea expedition. However, with the outbreak of war, the Dutch authorities in the East Indies

[1]Victor T. King, 'Introduction' to Carl Lumholtz, *Through Central Borneo*, Oxford University Press reprint, Singapore, 1991, pp. v–xviii.

could not guarantee Lumholtz's safety nor provide him with assistance in New Guinea. He therefore decided to explore Central Borneo. His expedition was Norwegian-sponsored, with some funding provided by British and Dutch learned societies and wealthy American and English patrons. From mid-1915 to mid-1916 Lumholtz travelled from the south coast of Borneo up the Barito River and then over the watershed to the Upper Mahakam. He then found his way downstream to the east coast at Samarinda. Lumholtz traversed the same regions as his fellow countryman, Carl Bock, had done in 1879–80. But Bock had started from the Mahakam and ended in the Barito, and Bock's love of the sensational contrasts with Lumholtz's sober descriptions.

Lumholtz demonstrates the cool, objective, measured tone of the professional explorer and authoritative scientist. Prior to his Borneo adventure he had already acquired an international reputation as an ethnographer, zoologist, and explorer, based on his scholarly publications on Australia and Central America. Unfortunately, Though Central Borneo *is the only substantial published record which Lumholtz has left us of his last expedition. He died in New York State in May 1922 from the recurrence of a tropical fever, which he had contracted some six years before in Borneo.*

This extract is taken from Chapters XX, XXI, and XXII of his book; it describes his arrival in the Upper Mahakam and his experiences among such Central Borneo peoples as the Penihings, Oma (Uma) Sulings, Long Glats, and the nomadic Punans and Bukats. Lumholtz gives us here a range of interesting material on Dayak character, material culture, everyday life, and ceremonies.

Unlike Boyle, Lumholtz is obviously wholly sympathetic of Dayak peoples and cultures, and it is instructive that in his book he uses the following prefaced quotation from Alfred Russel Wallace: 'We may safely affirm that the better specimens of savages are much superior to the lower examples of civilized peoples.'

Lumholtz was particularly skilled in communicating 'the personal, intellectual, and artistic qualities of native populations in a popular and interesting way to a wider educated readership' (King, 1991: xvi). Nevertheless, he still held to certain paternalistic and evolutionist assumptions about non-Western peoples.

A few minutes later we came in sight of the Mahakam River. At this point it is only forty to fifty metres wide, and the placid stream presented a fine view, with surrounding hills in the distance. In the region of the Upper Mahakam River, above the rapids, where we had now arrived, it is estimated there are living nearly 10,000 Dayaks of various tribes, recognised under the general name Bahau, which they also employ themselves, besides their tribal names.

The first European to enter the Mahakam district was the Dutch ethnologist, Doctor A. W. Nieuwenhuis at the end of the last century. He came from the West, and in addition to scientific research his mission was political, seeking by peaceful means to win the natives to Dutch allegiance. In this he succeeded, though not without difficulty and danger. Although he was considerate and generous, the Penihing chief Blarey, apprehensive of coming evil, twice tried to kill him, a fact of which the doctor probably was not aware at the time. Kwing Iran, the extraordinary Kayan chief, knew of it and evidently prevented the plan from being executed. Blarey did not like to have Europeans come to that country, which belonged to the natives, as he expressed it.

The Penihing kampong, Sungei Lobang, was soon reached. It is newly made, in accordance with the habit of the Dayaks to change the location of their villages every fourteen or fifteen years, and lies on a high bank, or rather a mud-ridge, which falls steeply down on all sides. It was the residence of the chief and the Penihings who brought us here, and if conditions proved favourable I was prepared to make a stay of several weeks in this populous kampong, which consists of several long, well-constructed buildings. The Dayaks assisted in putting up my tent, and of their own accord made a low palisade of bamboo sticks all around it as protection against the roaming pigs and dogs of the place. It proved of excellent service, also keeping away the obnoxious fowls, and during the remainder of my travels this measure of security, which I adopted, added considerably to my comfort. On receiving their payment in the evening the Dayaks went away in bad humour because they had expected that such a tuan besar [important person] as I was would give them more than the usual wages allowed when serving the Company, as the government is called. This tuan, they said, had plenty of money to boang (throw) away, and he had also a good heart.

Otherwise, however, these natives were kindly disposed and more attractive than either of the two tribes last visited. In husking rice the Penyahbongs, Saputans, and Penihings have the same method of gathering the grains back again under the pestle with the hands instead of with the feet, as is the custom of the Kenyahs and Kayans. All day there were brought for sale objects of ethnography, also beetles, animals, and birds. Two attractive young girls sold me their primitive necklaces, consisting of small pieces of the stalks of different plants, some of them odoriferous, threaded on a string. One girl insisted that I put hers on and wear it, the idea that it might serve any purpose other than to adorn the neck never occurring to them. Two men arrived from Nohacilat, a neighbouring kampong, to sell two pieces of aboriginal wearing apparel, a tunic and a skirt. Such articles are very plentiful down there, they said, and offered them at an astonishingly reasonable price.

Malay is not spoken here, and we got on as best we could—nevertheless the want of an interpreter was seriously felt. The chief himself spoke some and might have served fairly well, but he studiously remained away from me, and even took most of the men from the kampong to make prahus at another place. I was told that he was afraid of me, and certainly his behaviour was puzzling. Three months later I was enlightened on this point by the information that he had been arrested on account of the murder by spear of a woman and two men, a most unusual occurrence among Dayaks, who, as a rule, never kill any one in their own tribe. With the kampong well-nigh deserted, it soon became evident that nothing was to be gained by remaining and that I would better change the scene of my activities to Long Kai, another Penihing kampong further down the river.

A small garrison had been established there, and by sending a message we secured prahus and men, which enabled us to depart from our present encampment. There were some rapids to pass in which our collector of animals and birds nearly had his prahu swamped, and although it was filled with water, owing to his pluck nothing was lost. At Long Kai the lieutenant and Mr Loing put up a long shed of tent material, while I placed my tent near friendly trees, at the end of a broad piece of road on the river bank, far enough from the kampong to avoid its noises and near enough to the river to enjoy its pleasant murmur.

When going to their ladangs [farms] in the morning the Dayaks passed my tent, thence following the tiny affluent, Kai, from which the kampong received its name. Under the trees I often had interviews with the Penihings, and also with the nomadic Bukats and Punans who had formed settlements in the neighbouring country. Some of them came of their own accord, others were called by Tingang, the kapala [headman] of Long Kai, who did good service as interpreter, speaking Malay fairly well. From my tent I had a beautiful view of the river flowing between wooded hills, and the air was often laden with the same delicious fragrance from the bloom of a species of trees which I had observed on the Kasao River. Here, however, the odour lasted hours at a time, especially morning and evening. On the hills of the locality grow many sago palms, to which the natives resort in case rice is scarce.

It was quite agreeable to see a flag again, the symbol of the Dutch nation being hoisted every day on the hill where the military encampment was located, usually called benting (fortress). Even the striking of a bell every half-hour seemed acceptable as a reminder of civilisation. The soldiers were natives, mostly Javanese. The lieutenant, Th. F. J. Metsers, was an amiable and courteous man who loaned me Dutch newspapers, which, though naturally months out of date, nevertheless were much appreciated. We were about 1° north of equator and usually had beautiful, clear nights in the month of May. The Great Bear of the northern hemisphere was visible above the horizon and the planet Venus looked large and impressive. There were no mosquitoes and the air was fine, but at times the heat of the day was considerable, especially before showers. After two days of very warm weather without rain ominous dark clouds gathered in the west, and half an hour later we were in the thick of a downpour and mist which looked as if it might continue for days. But in inland Borneo one knows a rainstorm will soon belong to the past. Two hours later the storm abated and before sunset all was over, and the night came again clear and glorious.

One afternoon seven prahus with thirty-odd Dayaks were seen to arrive from down the river, poling their way. They were Kayans from Long Blu, en route for the Upper Kasao to gather rattan. Some of them called on me and evidently already knew of the expedition. They carried only rice as provisions and told me they intended to be away three months. On the Upper Kasao there is no more rubber to be found, and, according to them, on the upper part of Mahakam there is no more rattan.

The Penihings of Long Kai are good-natured and pleasant, and it was refreshing to be among real, natural people to whom it never occurs that nudity is cause for shame; whom the teaching of the Mohammedan Malays, of covering the upper body, has not yet reached. This unconsciousness of evil made even the old, hard-working women attractive. They were eager to sell me their wares and implements and hardly left me time to eat. Their houses had good galleries and were more spacious than one would suppose from a casual glance.

One morning I entered the rooms of one of the principal blians [shamans], from whom I wanted to buy his shield, used as a musical instrument to accompany his song. The shield looks like the ordinary variety used by all the tribes of the Mahakam and also in Southern Borneo, but has from four to ten rattan strings tied lengthwise on the back. In singing to call good spirits, antohs, especially in case somebody is ill, he constantly beats with a stick on one of the strings in a monotonous way without any change of time. Among the Penihings this shield is specially made for the blian's use, and unless it be new and unused he will not sell it, because the blood of sacrificial animals has been smeared on its surface and the patient would die. The only way I could secure one was by having it made for me, which a blian is quite willing to do.

This man paid little attention to my suggestion of buying, but suddenly, of his own accord, he seized the shield and played on it to show me how it was done. While he sings he keeps his head down behind the shield, which is held in upright position, and he strikes either with right or left hand. He had scarcely performed a minute when a change came over him. He stamped one foot violently upon the floor, ceased playing, and seemed to be in a kind of trance, but recovered himself quickly. A good antoh, one of several who possessed him, had returned to him after an absence and had entered through the top of his head. So strong is the force of auto-suggestion.

It was a matter of considerable interest to me to meet here representatives of two nomadic tribes of Borneo who had formed small settlements in this remote region. I had already made the acquaintance of the Punans in the Bulungan, but as they are very shy I welcomed the opportunity of meeting them on more familiar terms. For more than a generation a small number has been settled at Serratta, six hours walking 'distance from Long Kai. The other nomads, called Bukats, from the mountains around the headwaters of the Mahakam, have lately established themselves on the river a short distance above its junction with the Kasao; a few also live in the Penihing kampong Nuncilao. These recent converts from nomadic life still raise little paddi, depending mostly upon sago. Through the good offices of the Long Kai kapala people of both tribes were sent for and promptly answered the call. The Punan visitors had a kapala who also was a blian, and they had a female blian too, as had the Bukats.

The Punans are simple-minded, shy, and retiring people, and the other nomads even more so. The first-named are more attractive on account of their superior physique, their candid manners, and somewhat higher intellect. The natural food of both peoples is serpents, lizards, and all kinds of animals and birds, the crocodile and omen birds excepted. With the Bukats, rusa [deer] must not be eaten unless one has a child, but with the Punans it is permissible in any case. The meat of pig is often eaten when ten days old, and is preferred to that which is fresh. In this they share the taste of the Dayak tribes I have met, with the exception of the Long-Glats. I have known the odour from putrefying pork to be quite overpowering in a kampong, and still this meat is eaten without any ill effect. Salt is not used unless introduced by Malay traders. And evidently it was formerly not known to the Dayaks.

None of these jungle people steal and they do not lie, although children may do either. They were much afraid of being photographed and most of the Bukats declined. A Bukat woman had tears in her eyes as she stepped forward to be measured, but smiled happily when receiving her rewards of salt, tobacco, and a red handkerchief. It had been worth while to submit to the strange ways of the foreigner.

Both tribes are strictly monogamous and distinguished by the severe view they take of adultery, which, however, seldom occurs. While it is regarded as absolutely no detriment to a young girl to sleep with a young man, matrimonial unfaithfulness is relentlessly punished. Payment of damages is impossible. The injured Punan husband cuts the head from both wife and corespondent and retires to solitude, remaining away for a long time, up to two years. If the husband fails to punish, then the woman's brother must perform the duty of executioner. The Bukats are even more severe. The husband of an erring wife must kill her by cutting off her head, and it is incumbent on her brother to take the head of the husband. At present the Punans and Bukats are relinquishing these customs through fear of the Company.

The Bukats told me that they originally came from the river Blatei in Sarawak, and that Iban raids had had much to do with their movements. According to their reports the tribe had recently, at the invitation of the government, left the mountains and formed several kampongs in the western division. One of them, with short stubby fingers, had a broad Mongolian face and prominent cheek-bones, but not Mongolian eyes, reminding me somewhat of a Laplander.

The Punans and the Bukats have not yet learned to make prahus, but they are experts in the manufacture of sumpitans [blowpipes]. They are also clever at mat-making, the men bringing the rattan and the women making the mats. Cutting of the teeth is optional. The gall of the bear is used as medicine internally and externally. In case of fractured bones a crude bandage is made from bamboo sticks with leaves from a certain tree. For curing disease the Punans use strokes of the hand. Neither of these nomadic tribes allow a man present when a woman bears a child. After child-birth women abstain from work four days. When anybody dies the people flee, leaving the corpse to its fate.

Having accomplished as much as circumstances permitted, in the latter part of May we changed our encampment to Long Tjehan, the principal kampong of the Penihings, a little further down the river. On a favourable current the transfer was quickly accomplished. We were received by friendly natives, who came voluntarily to assist in putting up my tent, laying poles on the moist ground, on which the boxes were placed inside. They also made a palisade around it as they had seen it done in Long Kai, for the Dayaks are very adaptable people. Several men here had been to New Guinea and they expressed no desire to return, because there had been much work, and much beri-beri from which some of their comrades had died. One of them had assisted in bringing Doctor Lorenz back after his unfortunate fall down the ravine on Wilhelmina Top.

*   *   *

It is significant as to the relations of the tribes that not only Bukats and Punans, but also the Saputans, are invited to take part in a great triennial Bahau festival when given at Long Tjehan. Shortly after our arrival we were advised that this great feast, which here is called tasa and which lasts ten days, was to come off immediately at an Oma-Suling kampong, Long Pahangei, further down the river.

Though a journey there might be accomplished in one day, down with the current, three or four times as long would be required for the return. However, as another chance to see such a festival probably would not occur, I decided to go, leaving the sergeant, the soldier collector, and another soldier behind, and two days later we were preparing for departure in three prahus.

What with making light shelters against sun and rain, in Malay called atap, usually erected for long journeys, the placing of split bamboo sticks in the bottom of my prahu, and with the Penihings evidently unaccustomed to such work, it was eight o'clock before the start was made. Pani, a small tributary forming the boundary between the Penihings and the Kayans, was soon left behind and two hours later we passed Long Blu, the great Kayan kampong. The weather was superb and the current carried us swiftly along. The great Mahakam River presented several fine, extensive views, with hills on either side, thick white clouds moving slowly over the blue sky. As soon as we entered the country of the Oma-Suling it was pleasant to observe that the humble cottages of the ladangs had finely carved wooden ornaments standing out from each gable.

We arrived at Long Pahangei (h pronounced as Spanish jota) early in the afternoon. Gongs were sounding, but very few people were there, and no visitors at all, although this was the first day of the feast. This is a large kampong lying at the mouth of a tributary of the same name, and is the residence of a native district kapala. After I had searched every-where for a quiet spot he showed me a location in a clump of jungle along the river bank which, when cleared, made a suitable place for my tent. Our Penihings were all eager to help, some clearing the jungle, others bringing up the goods as well as cutting poles and bamboo sticks. Evidently they enjoyed the work, pitching into it with much gusto and inter-est. The result was a nice though limited camping place on a narrow ridge, and I gave each man one stick of tobacco as extra payment.

During our stay here much rain fell in steady downpours lasting a night or half a day. As the same condition existed higher up the river, at times the water rose menacingly near my tent, and for one night I had to move away. But rain in these tropics is never merciless, it seems to me. Back from the coast there is seldom any wind, and in the knowledge that at any time the clouds may give place to brilliant sunshine, it is not at all depressing. Of course it is better to avoid getting wet through, but when this occurs little concern is felt, because one's clothing dries so quickly.

The Oma-Sulings are pleasant to deal with, being bashful and unspoiled. The usual repulsive skin diseases are seldom seen, and the women are attractive. There appears to have been, and still is, much intercourse between the Oma-Sulings and their equally pleasant neighbours to the east, the Long-Glats. Many of the latter came to the feast and there is much intermarrying among the nobles of the two tribes. Lidju, my assistant and friend here, was a noble of the Long-Glats with the title of raja and married a sister of the great chief of the Oma-Sulings. She was the principal of the numerous female blians of the kampong, slender of figure, active both in her profession and in domestic affairs, and always very courte-ous. They had no children. Although he did not speak Malay very well, still, owing to his earnestness of purpose, Lidju was of considerable assistance to me.

The kampong consists of several long houses of the usual Dayak style, lying in a row and following the river course, but here they were separated into two groups with a brook winding its way to the river between them. Very large drums, nearly four metres long, hung on the wall of the galleries, six in one house, with the head somewhat higher than the other end. This instrument, slightly conical in shape, is formed from a log of fine-grained wood, light in colour, with a cover made from wild ox hide. An especially constructed iron tool driven by blows from a small club is used to hollow out the log, and the drum is usually completed in a single night, many men taking turns. In one part of the house lying furthest west lived Dayaks called Oma-Palo, who were reported to have been in this tribe a hundred years. They occupied 'eight doors', while further on, in quarters comprising 'five doors', dwelt Oma-Tepe, more recent arrivals; and both clans have married Oma-Suling women.

The purpose of the great feast that filled everybody's thoughts is to obtain many children, a plentiful harvest, good health, many pigs, and much fruit. A prominent Dayak said to me: 'If we did not have this feast there would not be many children; the paddi would not ripen well, or would fail; wild beasts would eat the fowls, and there would be no bananas or other fruits.' The first four days are chiefly taken up with preparations, the festival occurring on the fifth and sixth days. A place of worship adjoining the front of the easternmost house was being constructed, with a floor high above ground on a level with the gallery, with which it was connected by a couple of planks for a bridge. Although flimsily built, the structure was abundantly strong to support the combined weight of the eight female blians who at times performed therein. The hut, which was profusely decorated with long, hanging wood shavings, is called dangei and is an important adjunct of the feast, to which the same name is sometimes given. Ordinary people are not allowed to enter, though they may ascend the ladder, giving access to the gallery, in close proximity to the sanctuary.

Prior to the fifth day a progressive scale is observed in regard to food regulations, and after the sixth, when the festive high mark is reached, there is a corresponding decrease to normal. Only a little boiled rice is eaten the first day, but on the second, third, and fourth, rations are gradually increased by limited additions of toasted rice. The fifth and sixth days give occasion for indulgence in much rice and pork, the quantity being reduced on the seventh, when the remaining pork is finished. On the eighth and ninth days the regulations permit only boiled and toasted rice. Not much food remains on the tenth, when the menu reverts to boiled rice exclusively. Some kinds of fish may be eaten during the ten-day period, while others are prohibited.

It was interesting to observe what an important part the female blians or priest-doctors played at the festival. They were much in evidence and managed the ceremonies. The men of the profession kept in the background and hardly one was seen. During the feast they abstain from bathing for eight days, do not eat the meat of wild babi [pig], nor salt; and continence is the rule. Every day of the festival, morning, afternoon, and evening, a service is performed for imparting health and strength, called melah, of which the children appear to be the chief beneficiaries. Mothers bring babes in cradles on their backs, as well as their larger children. The blian, who must be female, seizing the mother's right hand with her left, repeatedly passes the blade of a big knife up her arm. The child in the cradle also stretches out its right arm to receive treatment, while other children and women place their right hands on the hand and arm of the first woman, five to ten individuals thus simultaneously receiving the passes which the blian dispenses from left to right. She accompanies the ceremony with murmured expressions suggesting removal from the body of all that is evil, with exhortations to improvement, etc.

This service concluded, a man standing in the background holding a shield with the inside uppermost, advances to the side of the mother and places it horizontally under the cradle, where it is rapidly moved forward and backward. Some of the men also presented themselves for treatment after the manner above described, and although the melah performance is usually reserved for this feast, it may be employed by the blian for nightly service in curing disease.

This was followed by a dance of the blians present, nine or ten in number, to the accompaniment of four gongs and one drum. They moved in single file, most of them making two steps and a slight turn to left, two steps and a slight turn to right, while others moved straight on. In this way they described a drawn-out circle, approaching an ellipse, sixteen times. After the dancing those who took part in the ceremonies ate toasted rice. Each day of the feast in the afternoon food was given to antoh by blians and girl pupils. Boiled rice, a small quantity of salt, some dried fish, and boiled fowl were wrapped in pieces of banana leaves, and two such small parcels were offered on each occasion.

Meantime the festive preparations continued. Many loads of bamboo were brought in, because much rice and much pork was to be cooked in these handy utensils provided by nature. Visitors were slowly but steadily arriving. On the fourth day came the principal man, the Raja Besar (great chief), who resides a little further up the river, accompanied by his family. The son of a Long-Glat father and an Oma-Suling mother, Ledjuli claimed to be raja not only of these tribes, but also of the Kayans. Next morning Raja Besar and his stately wife, of Oma-Suling nobility, accompanied by the kapala of the kampong and others, paid me a visit, presenting me with a long sugarcane, a somewhat rare product in these parts and considered a great delicacy, one large papaya, white onions, and bananas. In return I gave one cake of chocolate, two French tins of meat, one tin of boiled ham, and tobacco.

Domestic pigs, of which the kampong possessed over a hundred, at last began to come in from the outlying ladangs. One by one they were carried alive on the backs of men. The feet having first been tied together, the animal was enclosed in a coarse network of rattan or fibre. For the smaller specimens tiny, close-fitting bamboo boxes had been made, pointed at one end to accommodate the snout. The live bundles were deposited on the galleries, and on the fifth day they were lying in rows and heaps, sixty-six in number, awaiting their ultimate destiny. The festival was now about to begin in earnest and an air of expectancy was evident in the faces of the natives. After the performance of the melah and the dance of the blians, and these were a daily feature of the great occasion, a dance hitherto in vogue at night was danced in the afternoon. In this the people, in single file, moved very slowly with rythmic steps, describing a circle around three blians, including the principal one, who sat smoking in the centre, with some bamboo baskets near by. Next morning the circular dance was repeated, with the difference that the participants were holding on to a rope.

About four o'clock in the afternoon the Dayaks began to kill the pigs by cutting the artery of the neck. The animals, which were in surprisingly good condition, made little outcry. The livers were examined, and if found to be of bad omen were thrown away, but the pig itself is eaten in such cases, though a full-grown fowl or a tiny chicken only a few days old must be sacrificed in addition. The carcasses were freed from hair by fire in the usual way and afterward cleaned with the knife. The skin is eaten with the meat, which at night was cooked in bamboo. Outside, in front of the houses, rice cooking had been going on all day. In one row there were perhaps fifty bamboos, each stuffed with envelopes of banana leaves containing rice, the parcels being some thirty centimetres long and three wide.

During the night there was a grand banquet in all the houses. Lidju, my assistant, did not forget, on this day of plenty, to send my party generous gifts of fresh pork. To me he presented a fine small ham. As salt had been left behind we had to boil the meat à la Dayak in bamboo with very little water, which compensates for the absence of seasoning. A couple of men brought us two bamboos containing that gelatinous delicacy into which rice is transformed when cooked in this way. And, as if this were not enough, early next morning a procession arrived carrying food on two shields, the inside being turned upward. On these were parcels wrapped in banana leaves containing boiled rice, to which were tied large pieces of cooked pork. The first man to appear stepped up to a banana growing near, broke off a leaf which he put on the ground in front of me, and placed on it two bundles. The men were unable to speak Malay and immediately went away without making even a suggestion that they expected remuneration, as did the two who had given us rice. I had never seen them before.

The sixth day was one of general rejoicing. Food was exchanged between the two groups of houses and people were in a very joyful mood, eating pork, running about, and playing tricks on each other. Both men and women carried charcoal mixed with the fat of pork, with which they tried to smear the face and upper body of all whom they met. All were privileged to engage in this sport but the women were especially active, pursuing the men, who tried to avoid them, some taking refuge behind my tent. The women followed one man through the enclosure surrounding the tent, at my invitation, but they did not succeed in catching him. This practical joking was continued on the following days except the last.

The Oma-Palo had their own festival, which lasted only one day. It began in the afternoon of the sixth day and I went over to see it. The livers of the pigs were not in favourable condition, which caused much delay in the proceedings, and it was nearly five o'clock when they finally began to make a primitive dangei hut, all the material for which had been gathered. A few slim upright poles with human faces carved at the upper ends were placed so as to form the outline of a quadrangle. On the ground between them planks were laid, and on the two long sides of this space were raised bamboo stalks with leaves on, which leaned together and formed an airy cover. It was profusely adorned with wood shavings hung by the ends in long spirals, the whole arrangement forming a much simpler house of worship than the one described above. The kapala having sacrificed a tiny chicken, a man performed a war dance on the planks in superb fashion, and after that two female blians danced. Next morning I returned and asked permission to photograph the dancing. The kapala replied that if a photograph were made while they were working—that is to say, dancing—they would have to do all their work over again, otherwise some misfortune would come upon them, such as the falling of one of the bamboo stalks, which might kill somebody. Later, while they were eating, for example, there would be no objection to the accomplishment of my desire.

With the eighth day an increased degree of ceremonials became noticeable, and in order to keep pace therewith I was driven to continuous activity. On a muggy, warm morning I began work by photographing the Raja Besar, who had given me permission to take himself and his family. When I arrived at the house where he was staying he quickly made his preparations to 'look pleasant', removing the large rings he wore in the extended lobes of his ears and substituting a set of smaller ones, eight for each ear. He was also very particular in putting on correct apparel, whether to appear in warrior costume or as a private gentleman of the highest caste. His sword and the rest of his outfit, as might be expected, were of magnificent finish, the best of which Dayak handicraft is capable. He made altogether a splendid subject for the camera, but his family proved less satisfactory. I had to wait an hour and a half before his womenfolk were ready, femininity apparently being alike in this regard in all races. When they finally emerged from the house in great array (which showed Malay influence) they were a distinct disappointment.

The raja, who was extremely obliging, ordered the principal men of the kampong to appear in complete war outfit, and showed us how an imaginary attack of Iban head-hunters would be met. They came streaming one after another down the ladder, made the evolutions of a running attack in close formation, holding their large shields in front of them, then ran to the water and paddled away, standing in their prahus, to meet the supposed enemy in the utan [jungle] on the other side of the river.

At noon the female blians were preparing for an important ceremony in the dangei hut, with a dance round it on the ground later, and I therefore went up to the gallery. The eight performers held each other by the hands in a circle so large that it filled the hut. Constantly waving their arms backward and forward they moved round and round. Some relics from Apo Kayan were then brought in: a small, shining gong without a knob and a very large bracelet which looked as if it had been made of bamboo and was about eight centimetres in diameter. One of the blians placed the bracelet round her folded hands and then ran round the circle as well as through it; I believe this was repeated sixteen times. When she had finished running they all walked in single file over into the gallery in order to perform the inevitable melah.

Shortly afterward followed a unique performance of throwing rice, small bundles of which, wrapped in banana leaves, were lying in readiness on the floor. Some of the men caught them with such violence that the rice was spilled all about, and then they flipped the banana leaves at those who stood near. Some of the women had crawled up under the roof in anticipation of what was coming. After a few minutes passed thus, the eight blians seated themselves in the dangei hut and prepared food for antoh in the way described above, but on this occasion one of them pounded paddi with two short bamboo sticks, singing all the while.

A very amusing entertainment then began, consisting of wrestling by the young men, who were encouraged by the blians to take it up and entered the game with much enthusiasm, one or two pairs constantly dancing round and round until one became the victor. The participants of their own accord had divested themselves of their holiday chavats and put on small ones for wrestling. With the left hand the antagonist takes hold of the descending portion of the chavat in the back, while with the right he grasps the encircling chavat in front. They wrestled with much earnestness but no anger. When the game was continued the following morning the young men presented a sorry spectacle. Rain had fallen during the right, and the vanquished generally landed heavily on their backs in the mud-holes, the wrestlers joining in the general laugh at their expense. To encourage them I had promised every victor twenty cents, which added much to the interest.

Having concluded their task of feeding the antohs the blians climbed down the ladder and began a march in single file round the dangei hut, each carrying one of the implements of daily life: a spear, a small parang, an axe, an empty rattan bag in which the bamboos are enclosed when the woman fetches water, or in which vegetables, etc., are conveyed, and another bag of the same material suitable for transporting babi. Four of the women carried the small knife which is woman's special instrument, though also employed by the men. When the eight blians on this, the eighth day, had marched sixteen times around the dangei they ascended the ladder again. Shortly afterward a man standing on the gallery pushed over the flimsy place of worship—a signal that the end of the feast had come. On the previous day a few visitors had departed and others left daily.

The feast had brought together from other parts about 200 Oma-Sulings and Long-Glats. The women of both tribes showed strikingly fine manners, especially those belonging to the higher class, which was well represented. Some were expensively dressed, though in genuine barbaric fashion as indicated by the ornaments sewn upon their skirts, which consisted of hundreds of florins and ringits [Malay dollars]. It should be conceded, however, that with the innate artistic sense of the Dayaks, the coins, all scrupulously clean, had been employed to best advantage in pretty designs, and the damsels were strong enough to carry the extra burden.

The climax had been passed and little more was going on, the ninth day being given over to the amusement of daubing each other with black paste. On the tenth day they all went away to a small river in the neighbourhood, where they took their meals, cooking paddi in bamboo, also fish in the same manner. This proceeding is called nasam, and the pemali (tabu) is now all over. During the days immediately following

190

the people may go to the ladang, but are obliged to sleep in the kampong, and they must not undertake long journeys. When the feast ended the blians placed four eggs in the clefts of four upright bamboo sticks as sacrifice to antoh. Such eggs are gathered from hens that are sitting, and those which have become stale in unoccupied nests are also used. If there are not enough such eggs, fresh ones are taken.

*   *   *

Every night while we were camped here, and frequently in the day, as if controlled by magic, the numerous dogs belonging to the Dayaks suddenly began to howl in chorus. It is more ludicrous than disagreeable and is a phenomenon common to all kampongs, though I never before had experienced these manifestations in such regularity and perfection of concerted action. One or two howls are heard and immediately all canines of the kampong and neighbouring ladangs join, perhaps more than a hundred in one chorus. At a distance the noise resembles the acclamations of a vast crowd of people. The Penihings and Oma-Sulings treat man's faithful companion well, the former even with affection; and the dogs, which are of the usual type, yellowish in colour, with pointed muzzle, erect ears, and upstanding tail, are in fine condition. A trait peculiar to the Dayak variety is that he never barks at strangers, permitting them to walk on the galleries or even in the rooms without interference. Groups of these intelligent animals are always to be seen before the house and on the gallery, often in terrific fights among themselves, but never offensive to strangers.

They certainly serve the Dayaks well by holding the pig or other animal at bay until the men can come up and kill it with spear. Some of them are afraid of bear, others attack them. They are very eager to board the prahus when their owners depart to the ladangs, thinking that it means a chase of the wild pig. Equally eager are they to get into the room at night, or at any time when the owner has left them outside. Doors are cleverly opened by them, but when securely locked the dogs sometimes, in their impatience, gnaw holes in the lower part of the door which look like the work of rodents, though none that I saw was large enough to admit a canine of their size. One day a big live pig was brought in from the utan over the shoulder of a strong man, its legs tied together, and as a compliment to me the brute was tethered to a pole by one leg, while the dogs, about fifty, barked at and harassed it. This, I was told, is the way they formerly were trained. As in a bull-fight, so here my sympathy was naturally with the animal, which managed to bite a dog severely in the side and shook another vigorously by the tail. Finally some young boys gave it a merciful death with spears.

A woman blian died after an illness of five days, and the next forenoon a coffin was made from an old prahu. She had not been ill long, so the preparations for the funeral were brief. Early in the afternoon wailing was heard from the gallery, and a few minutes later the cortege emerged on its way to the river bank, taking a short cut over the slope between the trees, walking fast because they feared that if they lingered other people might become ill. There were only seven or eight members of the procession; most of whom acted as pall-bearers, and all were poor people. They deposited their burden on the bank, kneeling around it for a few minutes and crying mournfully. A hen had been killed at the house, but no food was offered to antoh at the place of embarkation, as had been expected by some of their neighbours.

Covered with a large white cloth, the coffin was hurriedly taken down from the embankment and placed in a prahu, which they immediately proceeded to paddle down-stream where the burial was to take place in the utan some distance away. The reddish-brown waters of the Mahakam, nearly always at flood, flowed swiftly between the walls of dark jungle on either side and shone in the early afternoon sun, under a pale-blue sky, with beautiful, small, distant white clouds. Three mourners remained behind, one man standing, gazing after the craft. Then, as the prahu, now very small to the eye, approached the distant bend of the river, in a few seconds to disappear from sight, the man who had been standing in deep reflection went down to the water followed by the two women, each of whom slipped off her only garment in their usual dexterous way, and all proceeded to bathe, thus washing away all odours or other effects of contact with the corpse, which might render them liable to attack from the antoh that had killed the woman blian.

In the first week of June we began our return journey

against the current, arriving in the afternoon at Data Lingei, an Oma-Suling kampong said to be inhabited also by Long-Glats and three other tribes. We were very welcome here. Although I told them I did not need a bamboo palisade round my tent for one night, these hospitable people, after putting up my tent, placed round it a fence of planks which chanced to be at hand. At dusk everything was in order and I took a walk through the kampong followed by a large crowd which had been present all the time.

Having told them to bring all the articles they wanted to sell, I quickly bought some good masks and a number of tail feathers from the rhinoceros hornbill, which are regarded as very valuable, being worn by the warriors in their rattan caps. All were 'in the market', prices were not at all exorbitant, and business progressed very briskly until nine o'clock, when I had made valuable additions, especially of masks, to my collections. The evening passed pleasantly and profitably to all concerned. I acquired a shield which, besides the conventionalised representation of a dog, exhibited a wild-looking picture of an antoh, a very common feature on Dayak shields. The first idea it suggests to civilised man is that its purpose is to terrify the enemy, but my informant laughed at this suggestion. It represents a good antoh who keeps the owner of the shield in vigorous health.

The kapala's house had at once attracted attention on account of the unusually beautiful carvings that extended from each gable, and which on a later occasion I photographed. These were long boards carved in artistic semblance of the powerful antoh called nagah, a benevolent spirit, but also a vindictive one. The two carvings together portrayed the same monster, the one showing its head and body, the other its tail. Before being placed on the gables a sacrifice had been offered and the carvings had been smeared with blood—in other words, to express the thought of the Dayak, as this antoh is very fierce when aroused to ire, it had first been given blood to eat, in order that it should not be angry with the owner of the house, but disposed to protect him from his enemies. While malevolent spirits do not associate with good ones, some which usually are beneficent at times may do harm, and among these is one, the nagah, that dominates the imagination of many Dayak tribes. It appears to be about the size of a rusa, and in form is a combination of the

Masks, from Henry Ling Roth, *The Natives of Sarawak and British North Borneo*, Truslove and Hanson, London, 1896.

body of that animal and serpent, the horned head having a disproportionately large dog's mouth. Being an antoh, and the greatest of all, it is invisible under ordinary conditions, but lives in rivers and underground caves, and it eats human beings.

Lidju, who accompanied me as interpreter and to be generally useful, had aroused the men early in the morning to cook their rice, so that we could start at seven o'clock, arriving in good time at the Kayan kampong, Long Blu. Here, on the north side of the river, was formerly a small military establishment, inhabited at present by a few Malay families, the only ones on the Mahakam River above the great kihams [rapids]. Accompanied by Lidju I crossed the river to see the great kampong of the Kayans.

Ascending the tall ladder which leads up to the kampong, we passed through long, deserted-looking galleries, and from one a woman hurriedly retired into a room. The inhabitants were at their ladangs, most of them four hours' travel from here. Arriving finally at the house of Kwing Iran, I was met by a handful of people gathered in its cheerless, half-dark gallery. On our return to a newly erected section of the kampong we met the intelligent kapala and a few men. Some large prahus were lying on land outside the house, bound for Long Iram, where the Kayans exchange rattan and rubber for salt and other commodities, but the start had been delayed because the moon, which was in its second quarter, was not favourable. These natives are reputed to have much wang [money], owing to the fact that formerly they supplied rice to the garrison, receiving one ringit for each tinful.

Though next day was rainy and the river high, making paddling hard work, we arrived in good time at Long Tjehan and found ourselves again among the Penihings. During the month I still remained here I made valuable ethnological collections and also acquired needed information concerning the meaning and use of the different objects, which is equally important. The chief difficulty was to find an interpreter, but an intelligent and efficient Penihing offered his services. He 'had been to Soerabaia', which means that he had been at hard labour, convicted of head-hunting, and during his term had acquired a sufficient knowledge of Malay to be able to serve me. My Penihing collections I believe are complete. Of curious interest are the many games for children, among them several varieties of what might be termed toy guns and different kinds of puzzles, some of wood while others are plaited from leaves or made of thread.

The kampong lies at the junction of the Mahakam and a small river called Tjehan, which, like several other affluents from the south, originates in the dividing range. The Tjehan contains two or three kihams but is easy to ascend, and at its head-waters the range presents no difficulties in crossing. This is not the case at the sources of the Blu, where the watershed is high and difficult to pass. Small parties of Malays occasionally cross over to the Mahakam at these points as well as at Pahangei. In the country surrounding the kampong are several limestone hills, the largest of which, Lung Karang, rises in the immediate vicinity.

Doctor Nieuwenhuis on his journey ascended some distance up the Tjehan tributary, and in the neighbourhood of

195

Lung Karang his native collector found an orchid which was named *phalænopsis gigantea*, and is known only from the single specimen in the botanical garden at Buitenzorg, Java. On a visit there my attention was drawn to the unusual size of its leaves and its white flowers. I then had an interview with the Javanese who found it, and decided that when I came to the locality I would try to secure some specimens of this unique plant. Having now arrived in the region, I decided to devote a few days to looking for the orchid and at the same time investigate a great Penihing burial cave which was found by my predecessor.

Accompanied by two of our soldiers and with five Dayak paddlers, I ascended the Tjehan as far as the first kiham, in the neighbourhood of which I presumed that the burial cave would be and where, therefore, according to the description given to me, the orchid should be found. There was no doubt that we were near a locality much dreaded by the natives; even before I gave a signal to land, one of the Penihings, recently a head-hunter, became hysterically uneasy. He was afraid of orang mati (dead men), he said, and if we were going to sleep near them he and his companions would be gone. The others were less perturbed, and when assured that I did not want anybody to help me look for the dead but for a rare plant, the agitated man, who was the leader, also became calm.

We landed, but the soldier who usually waited upon me could not be persuaded to accompany me. All the Javanese, Malays, and Chinamen are afraid of the dead, he said, and declined to go. Alone I climbed the steep mountain-side; the ascent was not much over a hundred metres, but I had to make my way between big blocks of hard limestone, vegetation being less dense than usual. It was about four o'clock in the afternoon when, from the top of a crest which I had reached, I suddenly discovered at no great distance, perhaps eighty metres in front of me, a large cave at the foot of a limestone hill. With the naked eye it was easy to distinguish a multitude of rough boxes piled in three tiers, and on top of all a great variety of implements and clothing which had been deposited there for the benefit of the dead. It made a strange impression in this apparently abandoned country where the dead are left in solitude, feared and shunned by their former associates.

No Penihing will go to the cave of the dead except to help carry a corpse, because many antohs are there who make people ill. The extreme silence was interrupted only once, by the defiant cry of an argus pheasant. As the weather was cloudy I decided to return here soon, by myself, in order to photograph and make closer inspection of the burial-place. I then descended to the prahu, and desiring to make camp at a sufficient distance to keep my men in a tranquil state of mind, we went about two kilometres down the river and found a convenient camping-place in the jungle.

On two later occasions I visited the cave and its surroundings, becoming thoroughly acquainted with the whole mountain. The Penihings have an easy access to this primeval tomb, a little further below, by means of a path leading from the river through a comparatively open forest. The corpse in its box is kept two to seven days in the house at the kampong; the body of a chief, which is honoured with a double box, remains ten days. According to an otherwise trustworthy Penihing informant, funeral customs vary in the different kampongs of the tribe, and generally the box is placed on a crude platform a metre above the ground.

As for the orchid, I, as well as the Dayaks, who were shown an illustration of it, searched in vain for three days. There is no doubt that I was at the place which had been described to me, but the plant must be extremely rare and probably was discovered accidentally 'near the water', as the native collector said, perhaps when he was resting.

Through Central Borneo. An Account of Two Years' Travel in the Land of the Head-Hunters between the Years 1913 and 1917, Charles Scribner's Sons, New York, 1920, pp. 210–42.

13
Rapids and River Travel in East Borneo

H. F. TILLEMA

Hendrik Freerk Tillema was born in southern Friesland, the Netherlands, in 1870. He trained in pharmaceutics in the Universities of Leiden and Groningen between 1889 and 1894. From 1896 to 1914 he worked in Semarang, north-central Java, as a pharmacist, and also became very wealthy as a manufacturer of bottled mineral water, tonics, lotions, and toiletries. He had become interested in issues of public health and hygiene while living in the East Indies, and having made his fortune he retired to Holland, devoting the rest of his life to research and writing. In particular, he worked tirelessly for the promotion of better living conditions for the native populations of the Indies. To this end, he undertook three major expeditions to the Indonesian archipelago in 1924–5, 1927–8, and 1931–3, which he funded himself, to familiarize himself with the lives and cultures of the different peoples there. His many publications on the East Indies were generally popular pieces of work, intended to help encourage the interest of ordinary Dutch men and women in the colonized peoples and to increase Dutch understanding of the everyday lives and problems of Indonesians.

His last expedition in 1932, when he was already over sixty years of age, was to the remote Apo Kayan area of Central Borneo. Tillema was a skilled photographer and film-maker; he took many striking images of his river journey into the interior and of the Kenyah and Punan peoples who lived there and whom he visited from his base at Long Nawang, the remote government outpost.

The following extract describes a part of Tillema's arduous and dangerous 44-day journey along the treacherous Kayan River from the east coast town of Tanjung Selor to the interior. Tillema's equipment and supplies were transported in one of the regular Dutch convoys which journeyed to Long Nawang. This convoy comprised 11 long-boats and 89 rowers. Tillema's account is probably one of the best we have on river travel through Borneo rapids.

Tillema recorded his experiences in a book published in Dutch in 1938, lavishly illustrated with 336 black-and-white

plates. This was his last major work before his death in
November 1952. The English translation of Tillema's book,
with some additional photographs and an introduction to
Tillema's life and work, was published in 1989.

AMONGST the arrangements for the equipment, I had
also booked a place on a cargo ship, to depart at the
beginning or middle of October 1931. But dark,
threatening clouds had appeared on the financial horizon.*
No one knew what heavy thunderstorms might develop, but
it was certain that thunder and lightning were in the air.
Good friends advised me against going so far from home in
such troubled times. It would be irresponsible, they thought.
But what ought I to do? I had made my preparations with
great care; I had consulted various authorities, explaining to
them the great difficulties and problems of the expedition
and meeting with sympathy on all sides, so what would
people say if I abandoned my plan? And would they again be
willing to be of service when circumstances returned to
normal? Would it not give the impression of a lack of determination if I stayed at home? And how long would the abnor-
mal situation in the world last? Years? And would my physical
and mental state still be up to such a hard journey by then?
My wife settled the matter. She said, 'I should go if I were
you. You will get enough good work for years again, and
perhaps it will all turn out well in the end.' So I decided to
go. I am happy that we took that decision then, for delay
would surely have led to cancellation. The financial situation
in the world did not turn out well! Everyone knows that.

I left Antwerp on the 19th of October on board the
'Tajandoen', the fine new motor vessel of the 'Nederland'
line. I had already made the trip several times and always
with fully laden ships. Now the boat was as good as half
empty. Those yawning spaces made a sombre sight. The ship
was easily capable of 17 knots, but because there was so little
cargo, we sailed 'economically', i.e. at 12 knots. The voyage
thus took rather a long time, which was no problem for me,
for I was in no hurry. A very roomy and smart cabin offered
every opportunity for study and relaxation through reading.
We received news of the world situation now and again

---

*Tillema is referring to the Great Depression of the early 1930s.

while *en route*, from which it appeared that there was no question of any improvement! Rather the reverse. Arriving in Java, I got the impression that there were only a few who appreciated the seriousness of the situation. 'For in the East, everything turns out right.'

I went on to Banjarmasin to consult the Resident and the military commandant, and, at the same time, to respond to a request from the 'Basle [Basel] Mission' to do some filming for them. After about a week, I left for Tanjung Selor [Tandjoeng Seilor], arriving on the 16th of December 1931. I stayed with Lt. van Zanten.He was the acting area commander, in charge of the preparations for the convoy to the Apo Kayan. This was no small task! The *perahu* had to be checked and repaired (a *perahu* which has made the journey is always leaky, has suffered much, and from underneath looks like a broom because the fibres of wood have come loose from being dragged over stones and rocks); rice had to be bought for the crew; dried fish, tobacco, matches, stores had to be brought in by K.P.M., sorted, repacked, inventoried; it meant working from morning to night. A few figures may show what the convoy entailed:

The 89 rowers received	f. 3,560.00
Their food cost, for rice	f. 235.14
fish	f 308.00
tobacco and matches	f. 130.50
salt	f. 28.48
1,000 paraffin cans for packing stores cost	f. 500.00
Total	f. 4,762.12

The money received by the Dayak rowers for their work was spent at Tanjung Selor in buying goods such as paraffin, iron bars for their *mandau*, points and hooks for the poles they used to push or pull the *perahu* along; tobacco, beads, cottons, silver braid, velvet, sequins for their wives and children; salt, etc. A similar convoy leaves each month, so the merchants (Chinese, Malays, and Indians) make respectable incomes with their activities. Since the small garrison at Tanjung Selor also meets its various needs in their *toko* (shops), the convoys are of great importance to the place. For a few days a month, there is some life about it. Those are the days when the preparations are made for the convoy. When it has left, an almost total calm reigns.

As well as the space in the *perahu* needed for the stores destined for the distant garrison, the food for the Dayaks, soldiers, and the convoy leader, there has also to be room for what the rowers are taking with them. Regulations have been made in this respect, but they are not applied very strictly. My cases and boxes had to go, too! And with every convoy, there is all the measuring and weighing, fitting and fretting, for space is tight. Things which are not absolutely necessary at Long Nawang are often sent on a month later. The question was how much room was there for my effects. After consultation with van Zanten it turned out that I could take everything with me except two trunks. I was thus obliged to leave behind the things I had no particular need of, such as white suits, my European winter suits, some of the films and filmpacks, chemicals, etc. I had already allowed for this, and so I had taken a goodly amount of camphor with me. I had to divide everything up again and repack it, and make a new inventory. You will perhaps think that my account is beginning to become tiresome, but I can assure you that it cost me much effort and trouble to take all my various belongings with me! I repeat that the essentials must be showed so that it is unnecessary to search for them and unpack them to be able to retrieve them *en route*. But you should also consider that one never knows in advance which things will be unnecessary! A normal traveller takes with him money, clothes, food, paper, ink, and pencil. That is all. I took hundreds of other things with me. Might I just draw to your attention the question of taking money with you to a place where there is no banking system, no post office or shop, where one can cash neither cheques nor bills of exchange?

The day of departure dawned. On the morning of the 29th of December 1931, all the goods were placed on the quay and divided up by weight and volume. They were stowed so that the centre of gravity of the *perahu* was low, and so that they lay correctly in the water fore and aft, and thus were horizontal. This was all with an eye to the rowing and steering. The Dayaks embarked, and we, the convoy commander and myself, took our leave, then the trumpet was blown, the *dayung* [paddles] plunged into the water, the boat left the quay, and contact was broken with the outside world—for many weeks and months!

There I sat. Through the good offices of Lt. van Zanten, in a broad *perahu*, on a bench with room to stretch my legs, one of my film cameras by me, and a shade over my head. This comfort was exceptional. We travelled from seven o'clock in the morning until about five o'clock, rowing upstream as close as possible to the bank where the current is at its weakest and the least trouble is experienced with eddies. For the first few days, the current is never strong, at least if the river is not 'banjired' [flooded]. The paddles splashed into the water with a regular beat, and now and then the Dayaks pulled the *perahu* along with long poles fitted with an iron hook, which were fixed into or round the branches under the trees, or sometimes they made use of punt-poles. The sun was shining, but we were travelling beneath the trees which stretched their branches far over the river, so we were not troubled. At about five o'clock, camp was pitched in the forest. The *perahu* were secured. No special measures were needed along this calm part of the river. A watch was set at night, though, since there is always a chance of a *banjir* [flood].

I was happy to leave my hard bench at the end of a long day, for I was having trouble with saddle soreness; unfamiliarity. One of the rowers brought me a few cans of water so that I could bathe. A few branches stuck in the ground served as clothes- and bathtowel-hangers. Bathing was no pleasure because of the presence of *agas* [sandflies] and leeches.

After six days we reached Long Pangian [Panggean] where there is a fixed camp, a relic of a former small military post. By the camp lived a Chinese who traded with the Dayaks. He showed me a set of rhinoceros horns and a set of nails. Every part of a rhinoceros is valuable: horns, nails, skin, and excrement. In a good year, a set of horns could bring in about f. 400, a set of nails f. 25. The Chinese buy them. They cut up the skin into little 'rosettes' to boil up for soup. From the horns and nails they make medicines which serve as aphrodisiacs. The Dayaks hunt rhinoceros with a smooth-bore gun. They aim between the shoulder-blades if possible, since the skin is thinner there. If the animal is not dead after being hit, they pursue it with spear and blowpipe. To that end, they will track the animal for days and even weeks through the forest, a strenuous task involving many hardships. This hunting is now banned, because the rhinoceros is one of the 'protected' species. I was told by the Chinese of a nice detail about determining whether the animal was male or female from the traces left on the trees in the forest; the male has two horns, a large and a small one, while the female has only one. The males break off the branch whose leaves they want to eat, by getting it between their two horns and turning their head. This causes the branch to break. The females cannot do this, and so have to bite the branch off. Whether this information is true or not, I do not know. The horns are cut off with the skin. There is no cavity inside, as there is with a cow's horn. The horn is one with the skin. That evening there was a heavy rainstorm at the camp. But I let the downpour pass over calmly in the safety of the fixed camp.

We were to have left the camp at seven next day. At about six, the convoy commander, Heer de Waard, came to tell me that we could not get away because there was a *banjir* and the water was rising. I had naturally noticed this, but thought nothing of it, being unfamiliar with the technique of sailing. So that was a disappointment. I had a good deal of reading matter in one of my many trunks, and I ferreted it out, not without difficulty. While reading, I looked at the swelling waters continually. I marked the water-level with a knife on the steps which led to the river. Again and again, they disappeared under water. At about five in the evening, I observed that the level was constant. By the morning, the water turned out to have fallen by a metre. 'If it goes down another metre, the branches will come out of the water, and we shall be able to get under them,' said de Waard, the convoy commander. It continued to fall, and we could leave. We left. My beautiful shelter had been knocked down. Sadly, I could use it no more. Since we had to travel under the trees, as I have already mentioned, the shelter would have caused problems when the water was so high, and would also have been too dangerous what with the currents, and eddies, and the manoeuvres. Progress was by no means fast against the fierce current, yet we made a few kilometres that day. And now a singular feature of my journeys in Borneo; on the return journey, my travelling companion, Heer Israël, the Assistant Resident at Bulungan [Boeloengan], asked me what I thought of the *kiham* [rapid] at Long Pangian, as the fixed camp was called. I answered that there was no *kiham* there! Or that I had not seen one! 'Had I been asleep for that stretch?' I told him that we had had the *banjir*. What did we find on the return journey, then? That there was a *kiham*, a *kiham* of some note, but which I had been unable to see because the water then had been about 7 or 8 metres higher than on the return trip, as was apparent from the notches on the stairs! We came up against numerous *sawing* (banks of stones). They could sometimes be partly surmounted with the *teken* (punt-poles about 3 metres long), but usually it could only be managed using tow-ropes pulled by dozens of men while others ranged along the sides, lifted the *perahu*. It is very hard work.That was allowed for in the feeding arrangements. If I am not mistaken the calorific value of the ration of rice plus dried fish is 3,500 per day, which agrees with the amount considered necessary for a labourer doing the heaviest work in non-tropical regions.

Palm, who travelled to the Apo Kayan in 1909–10, observed in his notes: 'It is impossible to imagine anything more beautiful than a *perahu* being poled over the shoals by Dayaks. Standing bolt upright in the long, narrow boat, they push on their poles, pulling them evenly out of the water, swaying forwards and immersing them again; they bend their lithe, muscular bodies forwards, then sink backwards, resting all their weight on the poles and pushing the *perahu* forward. They show great deftness in the slender boat. Sometimes (very seldom! T.) they stumble in their movement, but their lithe bodies can move so that they finish up all right, like a cat that always lands on its feet.'

I have watched this scene hundreds and hundreds of times from the front veranda of my little house at Long Nawang, and every time I enjoyed it.

When sailing in convoy, the convoy commander's *perahu* is always in the vanguard. The head Dayak accompanies him. Each *perahu* has a crew of eight, one of whom is the leader. He is one of the elders who has authority over the other rowers by virtue of his birth and experience. The commander's *perahu* sails in front to give direction. The technique of sailing, which demands an exceptional degree of experience, often necessitates crossing the river. Great care must be taken during this crossing (*pelawat*) to look out for rocks which may be invisible, especially in a *banjir* when the water is muddy. If a boat should strike an underwater rock, there is great danger of capsizing. The experts usually recognize the danger points from the currents, from the waves, and from the rippling of the water. When the lead *perahu* has safely reached the other side, the others follow. Drifting downstream during the crossing cannot be allowed to happen. Numerous *kiham* consist of a series, sometimes separated from each other by a mere few hundred metres. If one drifts too far downstream during the crossing when between two *kiham*, the *perahu* gets into the wild water, fills up and capsizes, with all the associated consequences. The fellows row like devils in this situation. Everyone knows exactly what to do. The rowers sense it! Palm relates that one of his Malayan orderlies, a very clever man who could row well, lent a hand. Suddenly, 'Ajen Amat' was heard (Don't, Amat, stop it). The rowers had felt that things would end up badly and that only a Malay could have that on his conscience, for it could never happen to a Dayak!

When travelling up rivers, one has to be very careful about the overhanging trees, for the current in the river is very capricious. There are eddies, counter-currents, and upwellings everywhere. If one of the sides of the *perahu* touches the slanting stem of a tree, it is pushed underneath it by the current, sticks fast, and turns over.

Arriving at a *kiham*, the commander gives the signal to stop. He and the *kepala kuli* [*kapala koelie*: the head Dayak] disembark, followed by the experienced Dayaks, to see the lie of the land, how the *kiham* should be taken, and what tactics should be followed. Because it is different at every water-level.

It happened once during the consultations to decide the best route, that one of the headmen discovered a nest of bees about 100 metres upstream in one of the overhanging trees, a few metres above the water. It would have been impossible not to disturb the nest, firmly attached to the branches, with the long hooked poles while going up the river. A legion of bees would then have descended furiously on the rowers, leading to great misfortune. The commander had of necessity to cross over between two rapids instead of going straight along the bank of the river under the trees. The whole rapid, in fact, consisted of a series of *kiham* separated from each other by only about a hundred metres, while great blocks of rock and heavy boulders lay all over the river, with savage whirlpools and waves between them, and currents flowing sometimes left, sometimes right, with many currents running now upstream, then suddenly back downstream. There was no question of being able to cross with the heavily laden *perahu*. The *perahu* had thus to be unloaded. But how? The river bank was steep. There was no space for all the many goods. And besides, the goods could not be left lying there while the *perahu* crossed over empty! There was a long deliberation. Finally, a plan was prepared after about an hour's discussion. The men returned to the *perahu*. One was largely unloaded and sixteen determined men embarked. The *perahu* set out from the bank. Sergeant de Waard, as the man responsible, with the headman to the front, and the best, most experienced steersman at the bow, and his equal at the stern. The other fellows rowed like devils. The *perahu* was continually threatening to smash itself to matchwood against a rock, but the two experienced steersmen always managed to prevent this by pushing it away. The *perahu* repeatedly changed direction. Then underwater rocks threatened to capsize it. Fortunately, they could be seen by the steersmen, since the water was clear. If the water is muddy, the danger of capsizing is still greater, since they cannot be seen. Currents and waves do give indications to an experienced man, but where immediate action is needed in the *kiham*, there is no time to consider. A little way downstream, the high, savage waves of the *kiham* were threatening. The *perahu* moved across between the two raging, foaming waves. It did not drift a single metre. In anxious suspense, we all—close on 100 men—followed the dancing boat. It came to the other bank. A loud cheer went up! The goods were unloaded. The empty boat returned. It was a splendid sight to see those fellows paddling and splashing, their *dayung* entering the water exactly together. The way had been shown; de Waard and his men had borne the risk of capsizing and drowning. I congratulated them heartily. Four *perahu* now followed in succession.

Then it was the turn of the fifth—mine, with its expensive cameras and stores. I asked de Waard not to unload the crates of apparatus and films, but to leave behind for the moment the other goods. I took sixteen men on board and off we went! The crossing was more than beautiful from the point of view of filming. I filmed away, every nerve and muscle tense. There was no time to think of danger. The Dayaks rowed like the very devil; a truly beautiful sight!

Meanwhile, the commander's *perahu* had again done the crossing with half a load and stopped on the other side. De Waard and the headman followed the movements of my *perahu* from the other bank. Suddenly, they threw themselves into the *kali* [river] with a long rattan in their hands. I saw them standing up to their chests in water. All at once, they slung the rattan towards the *perahu* and the men grabbed it. Men on the bank wrapped the other end around a tree. A jolt, and that end lay slack in the water. Straight away, a few chaps jumped into the water, grabbed the end and held it fast, which was successful since the high speed had been braked by the first mooring. My *perahu* had got across without mishap. And I had filmed the crossing! De Waard told me that my boat had drifted too far and had almost got into the *kiham* downstream; this despite my rowers having redoubled their efforts to prevent it. It was to stop this that de Waard had jumped into the water with the rattan. During the filming, I had noticed nothing of this. That I had dared to film was, I will be honest, thanks to the fact that I had not been conscious of the great danger. I had watched the *pelawat* so many times, but de Waard had never set out the dangers of it for me. He had thought it better not to worry me! The goods and the *perahu* all arrived safely on the opposite bank. De Waard's tactics were never to worry me in advance. Yes, he was a careful, competent convoy commander!

Now all the *perahu* had to be unloaded, since the full *perahu* were too heavy to be pulled through the *kiham* further upstream. At such a *kiham*, the river narrows with the result that the current is very strong there. 'Because of the fast current and the rugged shape of the banks, counter-currents are set up, causing whirlpools. The river then flows in all sorts of directions instead of one, so that it takes great attention and great effort to avoid the *perahu* breaking up on the rocks or filling with water. If the latter happens there is no holding it....

It makes a big difference whether the water is low or high. Some *kiham* are barely noticeable in low water, but in high water it foams and tosses from all sides, so that it demands the most extreme care and skill and strength to pass through!'

Palm, from whose notes I have borrowed this description, adds: 'It is impossible to give a proper picture of the rapids. The sight is overwhelming especially when there is a *banjir*. From the bank, the waves do not seem very high, but when one sits in a boat going down such rapids (He was going downstream. T.), one sees the wildly spinning waves projecting high above the banks ... and one is extremely uneasy.' I agree with Palm in this. Photos and films will contribute to giving a better picture of the rapids than is possible by description, but it is still 'ersatz'.

At the *kiham* of which I was speaking the *perahu* were unloaded. The goods had to be carried along the bank. There was no path. It was midday. The sun was burning vertically above our heads. There was not a breath of wind. It was hot enough to make you collapse. But the Dayaks cheerfully put the goods on their backs and leapt over the rocks and boulders with them! I now had even more respect for these fellows than I had had before over their labours in the whirlpools! But I had also to see that I got along the bank. I held myself fast with one hand to roots, shrubs, and branches (in the other, I was carrying my camera and film camera), climbing over smooth, often steep and pointed, glowing hot rocks, jumping over boulders, slowly blundering on, thinking now and again: 'The roads are a bit easier at Bloemendaal.'

The *perahu* were pulled through the *kiham* with great difficulty and then reloaded. The passage had taken half a day and we had progressed less than 1,000 metres! Finally, after hours and hours of slogging, after continuous care, exertion, and concentration, the whole fleet of boats had got through the series of rapids. It was then about three o'clock and we had been going since seven o'clock in the morning. De Waard had the camp set up, which would otherwise have been impossible before 5 or half past. But the men were tired ... and they had honestly earned a few hours rest, de Waard observed. It is a remarkable fact that the Dayaks always maintain their good humour, even when the work is hardest and the exertion greatest.

At *kiham*, one is always dependent on the state of the river. At each rapid—and there are dozens!—the convoy commander, on whom the responsibility rests, must consider: how can I get through, should I unload completely or partially, or can I get through with full boats? Are they too heavy?

To unload completely or even partially costs a lot of time. And with the convoy, time is an important factor. If the convoy takes rather long, then the commander has trouble from his superiors. If they have any idea of the difficulties, then there is no problem, but if it concerns someone who thinks that all he has to do is to teach the Dayaks how to paddle, how to hold their *dayung*, well then it becomes very hard for a young man to take decisions which depend on so many factors!

I was once watching as a half-full *perahu* was pulled through the *kiham*. Eight of them had already come through unscathed. My *perahu* as well. We lay waiting until I suddenly saw all the Dayaks running to the place where the *perahu* was being pulled up. What had happened? In the keel of the *perahu*—made from half a tree trunk—an 'eye' had been made when it was hollowed out. To it was fastened the tow-rope. And this 'eye' had broken with the strain of a 40-man pull, so that heavy waves had broken into the boat. As a precaution, the end of the rattan had also been made fast to the bow timbers. The Dayaks could hold the *perahu* for a little while and quickly unload it. Only one can of rice had gone missing. I do not know whether de Waard had to pay for it; probably not, for he was not in the least to blame for the accident.

In camp in the evenings, the day's journey accomplished without accident, the convoy commander was happy and cheerful. But at the beginning of the day, his responsibilities were oppressive, and cheerfulness was far away! I understood this perfectly and respected the dedication of the young man!

After eighteen days of heaving, pulling, clambering, and pushing, the convoy reached Busang [Boesang] Tinggang, the little camp at the beginning of the Brem [Bem-]-Brem rapids. *Perahu* cannot get through here. A path has been made alongside, 27 kilometres long. The *perahu* were unloaded and the goods carried along the path. This took about ten days. I found it pleasant to stay in the fixed camp, for sitting on such a hard bench in a rolling boat, with the continual driving to hurry up, the stumbling about over the hot, sharp rocks along the many *kiham*, the camps in the forest (with the nasty *agas* and leeches), however interesting it may have been, all had its unpleasant side, so it was nice to have a rest.

A Journey among the Peoples of Central Borneo in Word and Picture, edited and with an Introduction by Victor T. King, Oxford University Press, Singapore, 1989 (first published in 1938 as *Apo-Kajan. Een filmreis naar en door Centraal-Borneo*, van Munster's Uitgevers-Maatschappij, Amsterdam), pp. 48–61.

14
Official Tours in Colonial Sarawak

MALCOLM MacDONALD

From 1946 to 1948 Malcolm MacDonald was Governor-General of Malaya and British Borneo, and from 1948 to 1955 he served as Commissioner-General for the United Kingdom in South-East Asia. It was while he was performing his official duties in the region that he visited Sarawak several times, and Borneo People *provides us with a sympathetic record of his contacts with and experiences of the different populations there in the late 1940s and early 1950s. MacDonald travelled extensively in colonial Sarawak, paying his first visit there in July 1946. He describes the lives and cultures of the Land Dayaks, Ibans, Kayans, Kenyahs, Malays, Melanaus, and Chinese.*

Despite his wide experience of Sarawak, he had a special affection for the Rejang Ibans of the Kapit-Baleh region, and he became very close friends with the senior Iban of that area, Temonggong Koh. A substantial part of Borneo People *concerns the Ibans, and one extract below describes a visit to an Iban longhouse above the Pelagus Rapids on the Rejang River. In this he refers to a Bill Sochon: Major W. L. Sochon, DSO parachuted into Borneo in the later stages of the Japanese war to organize local forces against the enemy. This secret campaign also involved Tom Harrisson, who was air-dropped into another part of the island (see Passage 15, Tom Harrisson,* World Within*).*

MacDonald's visit to the Iban longhouse of Sandai in 1948 was undertaken in the company of the then Governor of Sarawak, Sir Charles Arden-Clarke, his private secretary, Bob Snellus, and the Resident of Sibu, Gordon Aikman.

The second extract describes MacDonald's visit to the Baram at the end of 1951. MacDonald had first journeyed along the Baram River, calling at Kayan and Kenyah longhouses, in August 1946. He had also met Tama Weng at the time, the great Kenyah leader of the Baram. In this later visit, MacDonald outlines some of the changes which had affected these native communities during the previous five years and describes his return to Tama Weng's home.

Although MacDonald was a British government officer on official visits, he saw himself continuing the traditions of the personal and benevolent rule of the Brookes. Therefore, it was his policy to travel to meet the people as regularly as possible. It so happened that MacDonald did not view these duties as a chore. He loved to spend his time in longhouses, and the warmth of the relationships which he forged with the people is reflected in the relaxed, humorous, intimate style which MacDonald uses to describe the natives of Sarawak.

THE rest of that day's trip was tranquil. The most exciting voyages through rapids were to take place on the morrow. Then our calm and poise would be severely tested; but in the meantime our temois glided on smooth waters, mile after mile and hour after hour, until we came to Penghulu Sandai's house. There we disembarked, for we were to spend the night under his roof.

Sandai was one of the chiefs who had greeted me with Temonggong Koh on both my visits to Kapit. In the latter stages of the war he showed conspicuous character, initiative and courage. His long-house stood at the farthest limit of Iban territory, close to the land of the Kayans, where Bill Sochon and another member of the parachute party, Sergeant Barrie, lay in hiding. The Kayans were preparing to fight in support of the British officers, but the attitude of the Ibans who controlled vast regions lower down the river was still uncertain. Much depended on their decision.

Sochon made contact with Sandai, the nearest Iban leader of consequence. They met privily, and the chief listened to the Englishman's tale. He gave cautious thought to the proposal that the Ibans should join an expedition to expel the Japanese, then said that he must consult his people, and left the meeting. He promised to give an answer later.

At that time a party of fourteen Japanese soldiers was

camped close to Sandai's house. He could easily betray Sochon and Barrie, and thereby gain a handsome reward. Many an Iban would have fallen to that temptation, and some of Sandai's followers with whom he discussed the situation were not averse to that course. But Sandai argued earnestly against them.

For two days Sochon and Barrie waited anxiously for his return. They did not know what his answer would be. On the next night they were wakened in their hiding-place beside the river by the [s]plash of oars. Peering through the bushes, they saw the silhouettes of men alighting stealthily from boats. The first to leap ashore was Sandai, betraying no sign whether he came as friend or foe. Over one shoulder he carried a heavily loaded sack.

Flinging a word of greeting to Sochon, he untied the sack and spilled its contents on the ground. In the darkness they looked like a heap of coconuts. Inspecting them, Sochon discovered that they were the heads of fourteen Japanese soldiers. That was Sandai's answer. On their way up-river to join the white men he and his braves had attacked the enemy garrison and killed every member of it.

We arrived at Sandai's house above the Pelagus Rapids in the late afternoon. He was waiting on the river bank to greet us. His pleasant face expressed a modest character, though he wore the full, proud panoply of an up-river Iban chieftain—on his head a tall plumed cap, in his ears heavy brass ear-rings, round his neck a huge oyster shell suspended on a necklace, over his shoulders a feathered war-cloak, round his loins finely patterned sirat [loincloth], and on his arms and legs bangles and garters.

A reception party of the principal men and women of his house stood at his side. As we landed an elder waved a cock over our heads and spoke friendly words of welcome. The bird expressed in unmistakable language its wish that we would go to an even hotter place than the Rejang on that sweltering afternoon. Then we climbed a muddy path to the long-house.

As we ascend the bank let me introduce the reader more adequately to the Ibans' homes, the famous long-houses of Borneo. No more authentic example of savage architecture exists in the world. At one time most pagan peoples of Sarawak lived in them. Nomadic tribes like the Punans never acquired such settled or splendid habitations, and, as I have already explained, many Land Dayaks abandoned long-houses when they fled for safety to mountain retreats. Recently the Melanaus, too, have deserted communal dwellings and adopted a style of individual family huts; but the Ibans, the Kayans, the Kenyahs and kindred peoples still live invariably in their traditional massive abodes. In some respects their building materials and architectural features differ from one another, but the general plan of their houses is similar. I shall describe a long-house belonging to Ibans as an example of the characteristics common to the homes of all long-house dwellers.

Usually a house is built beside a river. Since rivers are the only highways in Sarawak and the natives must travel and conduct trade by boat, it is convenient for them to live close to water. Their dwellings stand on cleared ground along the tops of the banks. A long-house is built of wood, is raised on piles, and stretches continuously for a remarkable distance over the ground. I can perhaps convey to the uninitiated reader an impression of its appearance by saying that it looks like an enormously elongated English barn on stilts. To enter it, one must climb a tree-trunk propped steeply from the ground to its doorway, with notches cut in the trunk to make rough steps. Sometimes a hand rail is provided, but often this aid is lacking and a visitor must trust to his powers of balance. The footholds are sometimes quite large, but even so, on a wet day the ascent is more like climbing a greasy pole than strolling up a staircase. I know to my cost, for once I slipped and fell in the mud.

This hazard illustrates the effectiveness of the elevation of the house as a means to achieve its original purpose, for long-houses were raised on stilts primarily for defence. In the bad old days in Sarawak it was wise to dwell a dozen feet above the earth. Then if enemies rushed the outer stockade round a house, they still could not easily enter the residence. Several yards of empty space yawned between the intruders and their intended victims. Prudent men constructed their home at a height which made it impossible for a foe to stand below the building and jab his spear through the floor.

The building's great length also is partly explained by requirements of security. A century ago Sarawak was a land where peace and order scarcely existed. Might was right, and the good things of the earth belonged to the strong. Every family's and every community's prime anxiety was for its safety, and all must be in a constant state of military pre-paredness. For this reason the Iban's home was literally his castle. The stronghold would be more formidable if it were not a small building containing one family, but a large place containing many families. The bigger the house, the stouter its walls. The more numerous its inmates, the more plentiful its guards. That was the simple explanation of the remarkable size of Bornean residences. It sheltered many households. Other reasons besides security may have played a part in producing this arrangement. Sociability, the instinct of primitive men to live in herds, the economic advantages of communal life, and other considerations perhaps influenced the decision; but the primary, compelling need was to create an effective fortress.

Insecurity of life and limb throughout Sarawak did not dis-appear automatically with the arrival of James Brooke. For the next two generations he and his successor gradually extended the area of peaceful government, but even during the reign of the third Rajah occasional head-hunting expedi-tions were conducted by bloodthirsty chiefs reluctant to abandon their ancestors' favourite sport. Almost until the present day, therefore, the original reason for building long-houses remained to some extent valid. Over the country as a whole, however, security has steadily increased during the last few decades, and mainly for this reason there has been a tendency for long-houses to be shorter. They now vary considerably in size. Small dwellings shelter about ten families, or approximately sixty persons. At the other extreme a few houses still accommodate almost a hundred families, or some six hundred inmates. Many are the homes of fifty or more families, but an average long-house probably accom-modates between twenty and thirty.

In fact, a long-house is not so much a house in our sense of the term, as a village. Every inhabitant of the community lives under one common, continuous roof. Except occasion-ally in the case of Kayans, long-houses are not grouped together. Each stands by itself, isolated. The nearest neigh-bouring residence is probably a few miles away, and may be many miles distant. In its own locality a long-house is the home of every living human being throughout the surround-ing countryside.

So one building may house anything up to six hundred people. It is, in that sense, like a large block of flats in a modern city. In such buildings, however, the apartments of the individual families are raised in storeys, one on top of the other. The structure is lifted upwards towards the sky, not stretched outwards along the ground. A long-house, on the other hand, extends horizontally over the earth. It is a one-storey building, a bungalow. Moreover, instead of the separate family apartments forming a compact group covering, say, a square area, they are set end to end in a continuous row. A large long-house may stretch for more than a quarter of a mile.

The dwelling's interior design is simple. Let the reader imagine that he has successfully climbed one of the tree-trunk stairways and reached the entrance to the house. In an Iban home on the Rejang he arrives first on an open platform running the whole length of the building. Crossing that, he steps into the dwelling proper. He will notice four elementary facts about the interior. First, the outer wall through which he has entered rises only two or three feet above floor level, the space between it and the roof being open except for posts supporting the rafters. Thus light and air enter the place. Second, a continuous wall stretches the length of the houses and divides it longitudinally into two more or less equal halves. Third, the half of the building in which he now stands is open from end to end, forming a long, uninterrupted gallery. Fourth, in the central wall dividing the house a succession of doors leads into the other half of the dwelling.

If he walks through one of these doors, he will discover on its other side a very different plan. He enters a room lit by a small window. The place contains an open fireplace and is usually sparsely furnished with the few belongings of an Iban family. Every door in the central wall of the house leads into such a chamber. The rooms are set side by side along the building, and each room is owned by one family. The number of these apartments in a long-house depends on the number of families living in it. If twenty families inhabit a house, there are twenty rooms; if forty-three families inhabit it, there are forty-three rooms, and so on.

The room is the private quarters of a family, and is used for sleeping, eating and gossiping *en famille*. Its inmates spend much of the rest of their time in the gallery outside. This gallery is, so to speak, a common-room for the whole populace. Everyone uses it. Indeed, it is much more than a common-room, for it fulfils many purposes. It is like a village street on which all the separate family apartments open, and along which the villagers walk when paying calls on friends. It is the village club where the lads and worthies of the place gather to exchange news, tell stories, talk scandal and imbibe drink. It is the village hall where concerts and dances are held, where public meetings of the community take place, and where distinguished visitors are ceremoniously received. It is, in fact, the hub and centre of social life in this communal society. At nights it also becomes a dormitory. While parents, grandparents, newly wed couples and children in an Iban family sleep in their private apartments, the young bachelors

'Interior of a Sea Dyak Long-house', from William T. Hornaday, *Two Years in the Jungle: The Experiences of a Hunter and Naturalist in India, Ceylon, The Malay Peninsula and Borneo*, Kegan Paul, Trench and Co., London, 1885.

are relegated to the public gallery outside, and the unmarried girls occupy a loft above.

To complete this account of an Iban house I need add only one further explanation. Along the front of the roof-covered gallery extends, as I have said, an open platform or veranda. Its main function is achieved at harvest time, when the crop of padi is spread there in the sunlight to dry.

Kayan and Kenyah houses are similar in principle to those of Ibans, but differ in important architectural details. The principal difference between the two types lies in their comparative sizes. Iban dwellings are usually shorter and lower than the others. The roof is less spacious, the gallery narrower and the rooms more cramped. Kayan and Kenyah architecture is conceived on a grander scale. Their houses are in many ways rough and primitive, for they are built by jungle people with a few simple implements; but their construction is a triumph of the builder's craft. They are splendid in their vastness, and have a strength which in a temperate climate would endure for a thousand years. In the equatorial forest, however, perpetual damp and voracious insects slowly but steadily devour their substance, and even the mightiest among them last effectively little more than a human generation.

Penghulu Sandai's house was comparatively small, being only half finished. The chief and his followers had moved into it a few months earlier. Previously they lived on the other side of the river; but an epidemic of sickness afflicted them there and persuaded them that evil spirits had placed a curse on their home. They resolved to flit, and the elders studied the omens to decide where their new dwelling should be. Observing the flights of spider-hunters and listening for the cries of trogons, they permitted these significant ornithological phenomena to point the general direction, and then the exact site was settled when the medicine man saw clearly in a dream the location of the present house.

Sandai and his friends at once began to build, and had completed seven family quarters with the gallery attached before work was suspended while they gathered their padi harvest. When that task was done the house would be extended to include several more apartments.

Although the place was comparatively small, the common-room running its length was already very spacious. When I entered it, I was reminded of a long gallery in a stately Elizabethan home in England, though its almost naked inmates contrasted sharply with the extravagantly over dressed, hooped, breeched and ruffed first inhabitants of those lordly mansions. We had arrived at an hour when the natives were returning from daily toil, when they liked to gather sociably in the gallery for gossip. The great chamber was gradually filling with people. Men strolled to and fro or squatted in talkative groups on the floor. A few women stood inquisitively by, with babes-in-arms cradled in scarves round the mothers' necks. Other ladies of the household were for the moment too busy to join the evening assembly, for they were preparing their families' and our suppers, cooking rice, roasting chicken and boiling eggs over fires in their private apartments. But they kept making shy appearances in the darkening gallery, flitting from one room into another where they wished to fetch things—or pretended to fetch things, so that they might have another excuse for a surreptitious glance at us white strangers. Mostly they wore only necklaces, belts and skirts; but the younger girls had donned all their finery.

This concourse of human beings did not complete the company. Countless dogs scampered through the gallery, sniffing unceremoniously at the walls and at each other, a pack of flea-infested mongrels kept by the Ibans for hunting deer and wild boar. Chickens strolled through the house, stepping delicately between the feet of their owners and cackling with fright when they misjudged a situation and received a kick in the tail. On either side of the door of Penghulu Sandai's room a pair of splendid cocks were tethered by their legs. Cock-fighting is a popular sport unlawfully enjoyed by the up-river people. Sandai's birds were handsome, colourful creatures, and they stood with a proud air, as if conscious of their aristocratic status among the domestic animals. Every now and then they drew attention to themselves by lusty, long-drawn 'cock-a-doodle-doos'.

This mixed gathering made a considerable noise, and to the conversation of men, yapping of dogs, crowing of cocks and clucking of hens was added also the grunting of pigs. The pigs did not actually live in the gallery, but stayed below its split-bamboo floor sniffing ceaselessly at the dirt which accumulates beneath an Iban house. Most of the household slops and refuse were disposed of by being poured or pushed through the floor to make a meal for the pigs.

Night fell quickly, and lights were soon lit in the gallery. Half a dozen brass lamps like tall candlesticks stood on the floor, spaced at equal distances along its length. Beside each collected a circle of squatting men, and some women also now began to settle in the shadows behind their husbands and sons. It was attractive, looking down the room and seeing the successive small points of light illuminating the vivid faces or accentuating the black silhouettes of people gathered round them. The men smoked and talked. The women stayed silent, puffing cheroots or chewing betel-nuts as they listened to their lords' words of wisdom or mirth. Aikman and I joined one group and engaged in conversation on local politics.

In the gloom of the gallery I perceived bamboo fish-traps and wooden mortars for pounding padi. A painted war-shield leaned against one pillar, and two or three parangs hung on the walls. A gong and a drum stood in one corner beside rolls of cane matting. For half its length the great apartment rose clear to the rafters, but the other half had a lower ceiling which formed the floor of an upper chamber. There the padi was stored, and there the unmarried girls slept at nights.

In an Iban house the bachelors sleep in the public gallery and, in some districts, the unwed damsels sleep in this loft. A wooden staircase leads to their dormitory, and up it after dark climb youths who are courting.

A young buck so inclined will approach the couch of the girl who attracts him. She sits within her mosquito net, perhaps expecting him. He crouches beside her and speaks of his love or discusses the weather, according to which tactic he thinks more prudent. If, after a while, she asks him to roll a cigarette or to wrap a betel-nut for her, he knows that he has her permission to stay the night. If, on the other hand, she asks him to go and poke the fire, or to perform some other task outside the bed, he knows that she does not desire him. Alternatively, she may light her small bed-lamp as a sign that he should go. He never questions her decision, nor seeks to overcome her reluctance. That would be bad manners. He leaves her, returns to the bachelors' quarters, and hopes for better luck next time.

A system of trial marriage is a recognized part of the Rejang

Ibans' social code. A youth may sleep with a girl a few times without particular importance being attached to their liaison. It is publicly regarded as an experiment between the pair, to see whether they would suit each other as husband and wife. But if a boy visits a girl several times and does not then propose marriage, her parents question him about his intentions. If he plans to wed her, they are satisfied. If he does not, he is forbidden to enter her mosquito net again.

After a brief trial period the young man or woman is still free to refuse marriage. This practice is a deliberate test of the compatability of their physical temperaments, the couple having presumably already decided that in other ways they would suit each other. A youth or girl may try several sleeping partners before he or she makes a final choice. When the resolve is made, the wedding is celebrated with traditional pagan ritual. At the climax of the ceremony a betel-nut is split in two, and one half is given to the bride and the other to the bridegroom. It is a symbol of a partnership in which they share all things together.

Ibans marry early, the girls often at fifteen or sixteen, and the youths when they are a year or two older. They are invariably monogamous, and after marriage are usually faithful to their mates. Divorce is not difficult, but is at least no more frequent than in 'civilized' societies. Generally the Bornean tribesmen are good husbands and devoted fathers.

When our dinner was cooked that evening, the Governor and our party ate from a table especially made for us in the gallery. Afterwards we squatted on the floor while the women of the house laid before us those rows of little dishes which serve to curry favour with the demons who haunt the Iban world. Dutifully we performed the whole ceremony of bedara, from throwing splashes of tuak over our shoulder for the spirits, to planting a cock's feathers on the pyramid of food arranged for their refreshment. The evil ones thus appeased, we turned to appeasing ourselves with rice-wine. As usual, girls came forward to fill and refill our glasses, and to sing a ditty before each drink.

Penghulu Sandai being a widower, his daughter Sri acted as our hostess. She attended upon me. In age she counted some twenty years and in looks she surpassed almost any native woman whom I had seen. She was tall and long-legged for an Iban, standing almost five and a half feet high.

Her oval face was handsome and her figure had the grace of a Greek statue. She moved with dignity, having a stately walk. Quiet and shy in manner, she nevertheless laughed quickly at a jest. At the moment she was unmarried, having lately divorced her husband. He must have been an idiot to allow himself to be separated from such a partner.

She knelt like a suppliant before me, poured a bumper of tuak and held it towards me while she sang a saga of countless verses. I thought that it would never end. Bob Snellus sat beside me and translated her words. She began, it seemed, by saying that she felt diffident in the presence of so renowned a visitor, but that she was overjoyed to welcome me to her home. My fame had spread up the Rejang after my first visit to Kapit, and the people of her long-house had been eager to see me. They were glad that this time I had come so far up-river. Then she resorted to outrageous flights of fancy, declaring that I possessed in high degree all the virtues of a great Iban. These she recounted in detail and at length, but at last she exhausted her list and handed me the drink. I quaffed it to the applause of the grinning head-hunters crouching around.

Meanwhile a buxom young lady called Luli performed a similar service for Arden-Clarke. A cousin of Sri, she was fashioned in coarser, plumper mould. Yet if her face was plain, it was also jolly, and with unaffected smiles she praised in song Arden-Clarke's manly virtues and gubernatorial gifts. Other girls regaled Aikman and the rest of our party with carols and tuak. Each solo was an essay in shameless flattery.

Sri and I conversed for a while, with Snellus as our interpreter. We discussed the prospects for the padi harvest, the nature of Sri's household duties, her taste in jewellery and trinkets, and other pleasant topics. Suddenly she filled my glass again and burst into fresh song. This time she told of the Iban gods, each one in turn, describing what they meant to her people. Snellus said that she would end by comparing me to the whole lot. Whether she did so or not I never learned, for his attention was distracted by a wild-eyed maiden who wandered up to him and began singing his praises in high falsetto.

Then a couple of minstrels started to strike their gongs, and another native gave a few hearty thumps on a drum. Everyone settled to enjoy some dancing. Sri shifted from her suppliant posture before me and reclined at my side, propping herself comfortably against a large, unused gong.

A man hung a hurricane lamp to the ceiling, where it cast a circle of light on the floor in front of us. A youth stepped from the outer darkness into this arena of illumination. He donned a war-cloak of hornbill's feathers and a cap crowned with Argus pheasant's plumes, strapped a parang round his waist and laid a shield upon the floor. Then he struck a statuesque pose, hesitating for a few moments before breaking into the first strutting steps of an Iban dance. With a sudden fierce yell and leap he began his performance.

The crowd of savages watched with the expert scrutiny of professional critics, such as sit in the stalls in London, Paris and New York on the first night of a new ballet. For the next two hours dancer followed dancer. Among them Sandai himself performed with gay, jaunty steps. He was happy that evening, smiling often at us and sometimes to himself. Whenever he glanced towards Sri his pride glowed in his face. He felt that she was doing the honours well.

Sri did not dance. She laughed at my urging that she should do so, declaring that she could not. But she told Luli and another girl to perform, and fetched some colourful sarongs for them to wear. They giggled, and the three retired to a room to dress up. When they emerged they were so splendidly attired that they looked like mannequins exhibiting a set of palace fashions. The orchestra started another throbbing tune, and Luli and her partner did a slow, restrained, arm-waving, foot-tapping women's dance.

When at length everyone who wished to jig had done so, the entertainment ended. Then the whole company gathered round the Governor and me, so that we could address them. In simple, friendly terms Arden-Clarke spoke of Sarawak affairs, and outlined the Government's policies. He described plans to instruct the natives in improved methods of agriculture, to bring health services to them by sending mobile dispensaries up-river, to build Iban schools and train Iban teachers, and to promote local self-government by the appointment of district councils. His audience listened carefully, made comments and asked questions. He answered them. The proceedings promised to be a good session of a long-house parliament.

Suddenly a distant roll of thunder sounded, and a moment

later rain began to patter on the roof. Before long a terrific storm was raging. Crash after crash of thunder broke above the house, a mighty wind howled round it, and torrents of rain fell on it like a tidal wave. Perhaps the evil spirits had taken umbrage at some defect in our offerings earlier in the evening, for they seemed determined to destroy us. The enraged elements made a tumultuous noise. Simultaneously the animals in the house took fright; dogs barked fearfully, cocks crowed vociferously and pigs squealed in terror. Unless we shouted at the tops of our voices, none of us mere human beings could make ourselves heard above the din. We roared with helpless laughter, and adjourned the meeting.

It was past midnight. Arden-Clarke and I went to bed in the penghulu's room, but not to sleep. The storm was one of those brief, impetuous, tropical affairs which soon subsides; but the excitement in the house was not so short-lived. The pack of mongrels whined periodically, the fighting-cocks raised their voices in lusty competition, and from beneath the floor came now and then the whimper of a pig suffering, it seemed, a nightmare. In the next room someone snored, from other apartments came the coughing of old men, and sometimes a baby cried in misery until it got comfort from its mother. Meanwhile outside our door Penghulu Sandai and his followers had resumed their political discussion. They argued keenly and loud, maintaining an incessant, noisy babble of talk. For an unconscionable time sleep was impossible, and after that it was fitful. Thus we passed a short night in a long-house.

It seemed only a few moments between the last throaty spasm of coughing which I heard before I fell asleep, and the first cockcrow that woke me in the morning. The dawn chorus of bird song in Sandai's home was a raucous outburst of crowing by chanticleers.

Arden-Clarke and I rose, put on bathing-trunks and walked down the path for a plunge in the river. When we arrived we found Sri and Luli in occupation of the jetty, having their morning baths. They were clad in work-a-day sarongs which clung sopping wet to their legs, for the girls had been immersed up to their necks in water. Now they sat on a raft of logs, douching their faces and soaping their bodies.

They smiled cheerily, waved hands to us and called, 'Tabeh, Tuans!'

THE BEST OF BORNEO TRAVEL

'Tabeh, Sri! Tabeh, Luli!' we answered.

They continued their toilet and we joined them. In the presence of so much splashing humanity the place seemed safe enough from crocodiles, so we dived into the river for a swim. The girls continued to lather and rinse themselves. They joked together, and laughed heartily when I slipped by mistake beneath the logs and nearly got drowned. When they had finished their ablutions, they filled gourds with water and carried them to the house, the first domestic duty of native women every day.

Arden-Clarke and I followed them shortly. Soon afterwards we travellers were ready to depart, and descended the path to go aboard our temois. The entire population of the house came to wave us goodbye.

*   *   *

More than two years passed before I travelled again along the Baram. In the meantime many changes had occurred on the great river. Only a few weeks after my previous visit Father Jansen died. The good old priest never spared his energy or strength, and suddenly his aged body could take the strain no longer. Gently he obeyed the summons of the intruder, Death.

Shortly afterwards Lallang also gave up the ghost and went to join her ancestors in whatever Elysian long-house they occupy in the next world. So when I journeyed up the Baram again towards the end of 1951 the river was like a landscape from which some familiar features had been removed.

Other landmarks, however, remained unchanged. John Gilbert was still the Resident at Miri, and once again he accompanied me. With us came also a long-legged, bespectacled, smiling American journalist, Howard Sochurek, who brought with him three cameras, two hundred rolls of film, a satchel full of lenses and filters, and various other photographic gadgets; for he was to take pictures of domestic life among the head-hunters for the edification of the readers of that well-known magazine, *Life*.

On our way up-river we took aboard our old friend Francis Drake, now the District Officer at Marudi. We also stopped at Long Ikang to embark Penghulu Gau. The young chief's house was the same neat dwelling that I remembered, and beside it now stood a more modern building with trim steps leading to its front door, spick-and-span whitewashed walls, and large European-style windows admitting light and air. It was the realization of one of Gau's dreams—a school raised by himself and his people, and daily filled with young-sters eager to learn reading, writing and arithmetic. Thus far had the local Kenyahs progressed since two years earlier, when only a couple of blackboards symbolized their passion for intellectual advancement.

The master of Long Ikang stood at the water's edge, awaiting our arrival. He climbed aboard our launch, his family waved goodbye to us, and we headed up-stream.

Gau's face was as expressive as ever, lined with wrinkles of thought and care and laughter. His charm of manner was positively courtly, and the assured dignity of his bearing was somehow enhanced by his semi-savage dress. Gilbert told me that the penghulu grew constantly more impressive. His zeal for introducing the benefits of modern knowledge among his people was increasingly infectious. Though frequently disap-pointed at the slow pace of progress, he patiently understood that the Government's limited financial resources and the Kenyah's conservative habits necessarily put a brake on the speed at which up-to-date public health measures, improved methods of agriculture and enlightened educational provi-sions could be effectively established among the long-house dwellers. He was learning what Sydney and Beatrice Webb used to call 'the inevitability of gradualness'.

His reforming influence was spreading steadily along the Baram and beyond. In addition to being a member of the District Council at Marudi and of the Divisional Council at Miri, he was now a member of the Council Negri at Kuching. In this he again outstripped his one-time mentor, the great chief Tama Weng Ajang, who occupied no seat on that supreme body. The situation had arisen with the older man's consent. Tama Weng lived so much farther up-river that the long journeys to Kuching would have involved him in a tedious loss of time. Moreover at certain seasons, when bad weather made the upper Baram unnavigable for days on end, he would sometimes have found it impossible to make the trip at all. So Penghulu Gau became the representative of the Kenyah people in the national council.

He was to come with us all the way to Tama Weng's house

at Long Akar, the ultimate Mecca of our present expedition. At last I could spare time to make the long journey to the great man's home, and I looked forward to meeting him in his seat of power. Our plan was to travel up-stream all day, every day for three days, resting the first two nights at long-houses on the way and reaching Long Akar before dusk on the third. We would stay with Tama Weng for the better part of twenty-four hours, and take another three days for the return voyage.

During the first day's journey above Marudi we went ashore at Long Lama, to sip tea with the Chinese towkays [merchants] while our baggage was transferred from our launch to the temois in which we would travel the rest of the way. Embarking in them, we soon passed the great mansion at Long Laput. Being pressed for time, we did not stop, knowing that if we tarried in that hospitable place we should not escape in a reasonable time. Besides, Lallang was dead. Without her the house would seem like a shell from which life had departed.

We knew, too, that its surviving inmates whom we wished to see, Kallang and Ubong, would greet us at Long Akar, where they were visiting Tama Weng. So we hastened forward. But I looked nostalgically upon Long Laput as we glided past. Its immense length was as impressive and almost unbelievable as ever. From its shadowy gallery a few residents stared at us, silent and listless, like mournful ghosts haunting a scene of former good fellowship. They seemed to be spectres from a dead past—and they did indeed represent an old order which was swiftly disappearing.

As we advanced upstream we saw signs of many changes creeping into Borneo's interior. The crews in the prahus which frequently passed us, for example, no longer appeared as splendid savages with bare, bronzed bodies glistening in the sunlight, but as brown men aping white men in coloured shirts, khaki trousers and felt hats.

We saw evidence of change, too, in the house at Murik, where we stayed the first night. A large and well-populated Kenyah dwelling, it bore many marks of a society in a state of transition in both material and spiritual affairs. The stairway climbing to its entrance was no longer a notched tree-trunk, but a broad, professionally carpentered flight of steps. Inside the building the central wall dividing the public gallery from the private apartments was smartly panelled in pseudo-Jacobean style, and its dozens of doors were furnished with modern handles and keyholes. In the rooms the window spaces were much larger than they used to be. In fact, much of the structure was clearly the work, not of the Kenyahs themselves, but of a Chinese contractor hired to provide the dwelling with up-to-date fittings.

Its inmates also showed signs of a significant change. Most of them had abandoned native dress. The men wore European cricket shirts and football trousers, while the women favoured Malay bajus [blouses] and sarongs. Only from the neck upwards were the ladies still Kenyahs, with saucy Baram hats, long black tresses and heavy brass ornaments hanging from mutilated ears.

Another innovation was remarkable. Above the doors of many family rooms hung carved wooden crosses, and several natives carried similar emblems on their own persons. A group of small girls, for instance, wore clusters of brass rings in their ears in traditional pagan style, while round their necks hung fine gold chains with golden crucifixes bearing the figure of Christ. Many of these Kenyahs were recent converts to Christianity.

We spent a quiet evening at Murik, with neither feasting nor dancing, as became a community preoccupied with the solemn business of changing its religion. Next morning we left early and made good progress up-river. All day our boats sped forward, and shortly before night we reached a house called Long Kiseh, where we would sleep.

It was a formidable edifice containing nearly fifty rooms. Unlike the house at Murik, it was a Kayan dwelling. Constructed by the inmates themselves, it had every customary detail of primitive, massive architecture; and just as there was no sign of a Chinese contractor's work in the building, so also there was no glimmer of a foreign religion among its inmates. They were still pagans from the tops of their heads to the tips of their toes, unredeemed and unrepentant in body and soul. The almost-naked men wore their hair shaven in front and long behind, while the women retained completely their Kayan form of dress. This community evidently felt no discontent with manners and beliefs which had satisfied their ancestors for centuries.

The contrast between one house and another on the

Baram at that moment of social and cultural transition was interesting. Some dwellings showed no trace of modifications in ancient customs; while others bore evidence of startling changes wherever you looked. Generally speaking the Kayan homes were strongholds of conservative tradition and the Kenyah houses were pioneers of new ideas. It was eloquent testimony to the influence of Tama Weng and Gau.

After supper at Long Kiseh a space in the gallery was cleared for dancing, and an audience of several hundred crouched round the stage. The dancers performed in a delightfully free spirit. No inhibitions appeared to check indulgence in time-honoured hilarity. The people's high spirits were spontaneous, for there was little drinking; but the place seemed to be full of clowns and humorists. Among the skits in their programme was a witty burlesque of a lugubrious Punan jig, and another was an unexpurgated edition of the notorious Monkey Dance. Then half a dozen old crones stepped the hallowed quadrille which in olden days was performed when new heads came into a house. As young belles forty years earlier they had perhaps executed it holding in their hands the fresh, bloody heads of enemies; but now they twirled for our amusement antique skulls wrapped in banana leaves. The aged harpies' limbs lacked the lithe maidenish grace of their girlhood, but still they were light of foot and dignified in carriage; and their gentle, elegant movements made the dance a spectacle of strange, melancholy, haunting beauty.

After first light next morning we left Long Kiseh, for we must lose no time in pressing forward. The Baram wound ceaselessly ahead of us. Along its course were many treacherous passages, which our boats negotiated with a mixture of caution and boldness born of long experience of the fitful whims of Borneo waters. We were travelling beyond the foothills into the true highlands. The river-bed cut deep into the ground, and lofty landscapes loomed everywhere around us. In some places the channel was so confined and the walls of hills so precipitous that we passed through literal gorges; but in others the slopes receded more gradually from the water's edge, and a wider variety of mountain scenery met our gaze.

When we started, patches of early morning mist hung over the earth. Here wisps of vapour trailed along narrow valleys, with hills protruding above them like huge boulders suspended in mid-air; and there white, fleecy clouds avoided the low ground and wreathed themselves instead round the mountain peaks. So sometimes the airy shapes of fog sprawled underneath dark masses of earth superimposed above them, and at other times that order was reversed. The density of different cloud forms varied as much as their altitudes. In some spots the mist was so opaque that all terrestrial matter behind it was hidden; but in others it was so transparent that the hills beyond were partially revealed. According to the degree of its translucence, the outlines of slopes were silhouetted darkly or faintly, like objects seen through veils of varying thicknesses. So the landscape appeared like the half-actual and half-ghostly, beautiful and mysterious world portrayed in classical Chinese paintings.

Later the mist dissolved, but always the heavens were overcast with a threat of rain. Sometimes a rift appeared in the clouds, and for a while the sun broke through. Then steep hillsides intruding close upon the river stood brilliantly exposed, their tropical vegetation luxurious, with every twig and leaf etched precisely by the sunlight.

At last, towards evening, we rounded a bend and spied on a distant promontory the white fort at Long Akar. We had arrived, and sped over that last stretch of water with the relief of a marathon runner who, after a testing race, exerts his final effort to dash home.

When we came to land, Tama Weng Ajang stood there to greet us. He was now Temonggong Tama Weng Ajang, having been elevated to the rank of temonggong a year earlier, thus sharing with the illustrious Iban, Koh, the distinction of the highest native title in the Government's gift.

He appeared as chubby and benevolent as ever, and was clad in a strange mixture of Oriental and Occidental, pagan and Christian styles which perhaps reflected accurately the rather confused state of his mind. On his head perched a Kenyah hat topped with hornbills' feathers, his hair was trimmed and combed in native fashion, and from his ears hung leopard-cat's teeth and brass ear-rings. A European blue pinstripe jacket, an American white singlet and a pair of khaki shorts covered the massive middle sections of his body, while below them his legs and feet were bare except for a pair of Kenyah garters. So at his two extremities—from the neck upwards and from the knees downwards—he looked like an unregenerate Bornean, but the intervening portions of his body might have been outfitted by an assortment of haberdashers from the Western world.

His house was situated about three miles above Long Akar, at a spot called Long San; so as soon as possible we re-embarked in our boats for that last lap of our journey. In the brief twilight which precedes the tropical night we arrived at Tama Weng's home. When our outboard motors fell silent a sound of rushing water filled the air, as if lively falls splashed nearby. In the half-dark I perceived the source of this sweet snatch of nature's music, the leaping white forms of breaking waves and flying spray among innumerable shadowy boulders. The whole width of the Baram above our landing stage was a seething mass of rapids, and all night and all day their lisping, bubbling murmur reminded the inmates of Long San of their proximity to the wild.

Along a pathway leading to their house stood rows of Kenyah notables, marshalled to give us friendly greeting. I was astonished at the neat Western clothing of the men and the beautiful modern dresses of the women. I had expected to see a degree of sophistication in Tama Weng's household, but not to meet a company of people clad almost as smartly as a well-groomed party in Singapore.

In the gloaming out of doors I gained a fleeting impression of their sartorial elegance; and when I saw it afterwards in the glare of lamplight indoors my astonishment at its splendour was heightened. Perhaps there was nothing very remarkable about the style of the men's attire, which consisted uniformly of Kenyah hair cuts, necklaces of yellow beads, variously coloured vests or shirts, blue shorts and native garters. What was surprising was their spotless cleanliness. They betrayed none of the casual dirt which invariably, and even unavoidably, gathered as a stain here or a splash there on the garb of other long-house dwellers. Not only were the faces of these Kenyahs perfectly washed and their hair immaculately brushed, but their shirts and trousers looked as if they had just returned from a high-class laundry.

The same quality marked the appearance of the women, but their gowns were distinguished also by unusual style. It was a sign of their shrewd taste that they did not discard their traditional hats, chic little bits of nonsense contrived of wisps

230

of grass, strips of animal fur, patches of coloured beads and gay birds' feathers which would cause a sensation if they appeared on the race-course at Ascot, in a mannequin parade in Paris or at a theatrical garden party in New York. Their hair also the ladies wore in native styles. They must have been aware of the contemporary popularity of permanent waves, Grecian curls, Eton crops and other beauty-parlour conceits; but they scorned this modern urge to turn women's hair from a lovely natural product into a highly manufactured article. In this, too, their judgment was sound, for the most glorious adornment of Bornean females is their magnificent long, loose, glossy tresses, as black as a raven's wings.

The females' elongated ear-lobes with clusters of heavy ear-rings were in more doubtful taste; but they were a dramatic, savage touch, and without them the most distinguishing feature of Kenyah and Kayan women would have been absent. Some of the younger girls, however, lacked this trait, their ears retaining natural shapes and sizes. These maids had grown up since new ideas crept up the Baram, and their parents decided not to subject them to the painful distortion. Yet most of the damsels concerned regretted the fact, feeling odd freaks with features different from their friends'. Almost they thought that their natural ears, not the disfigured ones of the others, were the deformed objects!

From the neck downwards the ladies at Lang San had abandoned customary tribal costume. Bare shoulders and long Kenyah robes had disappeared, gone the way of blow-pipes and poisoned arrows, banished to the limbo of discarded antiquities. In their place a new fashion was popular, based on the Malaysian baju and sarong. But whereas among many Malay women such articles are somewhat shapelessly made, fitting the body loosely, these Kenyah beauties were particular about their cut. The blouses fitted their figures with sleek precision, following obediently every graceful curve; and the skirts, of brilliant colours, hung in elegant folds from their waists to their feet. The natural taste of these allegedly wild jungle women showed artistic genius.

Observing them, I remembered the report on my previous visit to the Baram that, though Tama Weng favoured many reforms in Kenyah society, he insisted on his people retaining their traditional garb. I wondered what had happened to modify his earlier resolution, and he confided to me later that he strongly disapproved of the women's partial abandonment of Kenyah dress. He had done his best to prevent it, telling them that they should abjure foreign styles and maintain unspoiled the beautiful costume which their mothers, grandmothers and female ancestors through countless generations had favoured. The ladies flatly refused to listen to him. They had a notion—which many other women in various countries seem to share—that the robe of yesterday is a discreditable garment, a deplorable throw-back to some dark, uncultured age, as outmoded as a fig-leaf. Even the pleas of their husbands added to the orders of their chief could not uproot this little piece of femininity from their minds. No doubt Tama Weng's own sartorial foibles enabled them to destroy his argument with a fatal retort, for they could point out that his blue pinstripe jacket, white vest and khaki trousers would scarcely have found favour with his father, grandfathers and male ancestors through countless generations.

He told me, however, that he would insist on the ladies preserving their small Kenyah hats and their large Kenyah ears.

After greeting his wife on the river bank, and shaking hands with all the leading personalities there, we walked to the dwelling. Climbing a flight of well-carpentered steps, we entered a sumptuous apartment. This was not the chief's permanent home. He was in process of erecting a new house, and the place which we now visited was only a temporary lodging to serve until the other was finished. The future edifice stood close by, and before complete darkness fell we went to inspect it.

Its massive posts, beams and struts stood already in position, not yet covered with walls and floors and roof. They formed the colossal inner framework of a long-house, like the gaunt, intricate skeleton of an oversized prehistoric monster in a geological museum. On the ground around lay a scattering of scores of poles, hundreds of planks and thousands of shingles, all the bits and pieces required to finish the structure. Their craftsmanship showed excellent quality, and had a modern style. Clearly many of them had been supplied from a professional builder's yard.

The erection of the house was at that time the principal occupation of its inmates-to-be. In the meantime they lived in a smaller, temporary structure made of slim poles, narrow planks and light attap. Yet their pride in creation had inspired its making, too. Its main communal chamber, into which Tama Weng now led us, was large and high, made artistic by carved pillars and comfortable by benches, stools and small tables. My attention was at once attracted to its principal feature, a Christian altar in the centre of one wall. On it stood a massive wooden crucifix, and behind it was the most astonishing reredos that I have ever seen. In the place of sacred honour among its assortment of pictures hung a sad-eyed portrait of Christ, cut from the Christmas number of some Catholic magazine. On either side were coloured photographs of His Majesty King George VI and Her Majesty the Queen, Their Royal Highnesses Princess Elizabeth (as she then was) and the Duke of Edinburgh, and Generalissimo Chiang Kai-shek. The regal British couples were posed in the peaceful gilded apartments of Buckingham Palace, but the Chinese leader was represented directing an assault by his troops on a field of battle. Companies of infantry charged an enemy, while batteries of guns fired shells and squadrons of aircraft dropped bombs all round him.

This impressive evidence of Christian piety did not mean that Tama Weng had been baptized. He still wished to be, but I was told that he had not yet attained a sufficient state of grace. The religious instruction which Father Jansen started was being continued by the old man's successor, Father Bruggermann, who joined us later that evening. He said that the chief was making good progress in his studies, and that probably his baptism, with that of most of his followers, would not be long delayed. Some members of the house had already been christened.

A Catholic mission was now established at Long Akar, and Father Bruggermann resided permanently there. He managed a school and encouraged among his parishioners clean habits, the wise care of health, a progressive outlook in agriculture, and other reforming tendencies, as well as the deeper Christian way of life. The priest was popular, and exerted considerable influence over his flock.

I sensed a remarkable spirit animating Tama Weng's household. The community possessed an *esprit de corps* which I had not previously met in any long-house. They seemed conscious that they were a significant people, a progressive society, a band of pioneers achieving some historic purpose. Every man, woman and even child appeared to be imbued with this idea. Their feeling of self-importance was expressed not only in cleanly dress, but also in confident bearing, unaffected good manners, a natural approach to us government officials, and obvious pride in themselves and their home. The impression that they were a people of destiny was deliberately fostered by their chieftain, who moved among them with quiet, dignified, massive authority.

How different was all this from the careless, aimless, soulless atmosphere of the other great house at Long Laput!

Borneo People, Jonathan Cape, London, 1956, pp. 101–13, 279–88.

15
Stones and Kelabits in Interior Borneo

TOM HARRISSON

No book of Borneo travel would be complete without a contribution from Tom Harrisson: explorer, traveller, adventurer, naturalist, ethnologist, and archaeologist. He first went on a Borneo expedition in 1932 as a young man of twenty-one (see Passage 1 by Patrick Synge). But perhaps his best known and loved work on Borneo is World Within. *It was primarily based on his experiences during the last months of the Pacific War, when, as part of a secret Allied campaign, Major Harrisson was parachuted into Central Borneo to help organize native resistance against the Japanese.*

In an intimate portrait, Harrisson describes for us the everyday life and culture of the Kelabit peoples of Bario as they were in the mid-1940s. It is clear that he had a particular affection for the Kelabits. Harrisson tramped over much of Sarawak and some of the neighbouring territories during his long residence in Borneo, but his 'story' about the Kelabits has a special and personal quality. World Within *was written and published while Harrisson was Curator and Government Ethnologist at the Sarawak Museum, a post which he occupied from 1947 to 1966.*

The book is preoccupied with Kelabit traditions and the recent changes affecting these. The following extracts have been chosen to illustrate the particular style of the Kelabit longhouse, and beliefs and practices surrounding stone and petrification. Although legends and stories about 'turning to stone' are common in Borneo, it is among the Kelabits that stone and its cultural uses are given special expression.

BY the latter half of 1944 the Japanese had been in occupation of all Borneo—and South-East Asia—three years. By 1944 the first glow of a new order had faded; the signs that this new mastery was only temporary began to multiply rapidly. By the end of 1944 the 'Greater Co-Prosperity' regime was visibly foundering. Before the end of the following year it had been overthrown by force of arms and atoms. It was the singular fortune of the present writer to be the first visible sign, and in some ways symbol, of this transition, reversion or progression, in one of the remotest and until then least known parts of tropical Asia, a part of the world so far within it as to live—up until then—in a sense nearly outside it. Because of the Japs, in this latter part of 1944 the sweet soaring cry of the gibbons, black, white, swift and smart swinging against the canopy green; the faint singing of old ladies making mats by flickering gum-candlelight; and the echoing murmur of wind sniffling out of the cold, mist-laden mountain cliffs down onto the plain below; these and many, many, other noises (tree crash, cicada buzz, mongoose chuckle, the whistle of the blood-red and black hill partridge, grasshoppers, a million moving termites, piglets, bat swing, goat laugh, eagle owl and the legendary noises of the enspirited night—to name a few others) were, for the first time in far upland history, swamped for a few moments by the sound of a great mechanical device.

Lying in the bomb-aimer's blister of an American four-engined Liberator little was to be seen on this first flight. In fact the navigational plot between below and the existing map made, showed us nearly fifty miles out at nowhere. Meanwhile, scarcely dreaming and certainly feeling nothing of the land under our belly, from the clouds something very special was being cooked up for the Kelabits, their dolmens and their dragon jars down below. I was the (unwitting) chef.

If what I am going to try and describe is worth describing

in any sort of detail, it is first needful to have a clear idea of what those people down there, set in that monster mountain tangle, were caring about, trying for, fearing, loving, ignoring or avoiding *then*. Without some appreciation of goings-on inside their long-houses or within the vigorous bodies of these tall, strong men and thickset jolly women, anything that follows might be too misleading. But in giving such a picture I am forced to fall back largely upon what I learnt later. For I arrived among these people literally on the bones of my arse; and knowing nil. Indeed, at first they knew more about me, even if only mistakenly. In what follows, therefore, I have first tried to give an accurate picture of the Kelabit way of life at one crucial point in its evolution, developing in a particular kind of isolation, and about to be subject to something of a revolution (which is the subject matter of the rest of the book).

Thus, much of what happened in the inside of Borneo within the year may get a sort of interest and even significance as part of a long and infinitely intricate story— rather than simply a hotch-potch splash of this and that. Anyway, I cannot tell all this story in one book or life, only begin to. If there is a failure here to balance two rather different approaches to living, mine and 'theirs', I hope it may be forgiven me. After all, I already have some seventy thick notebooks full of elaborate field observations. No doubt, if I am spared from cirrhosis and the other occupational diseases of the colonial civil servant in the East, the time will come to produce the necessary volumes of apparent scholarship. Until that grinding day, this is no more than an attempt to reconstruct a post-megalithic chapter or two in the history of a little place and a few people including, for better or for worse, the present writer on this subject, within it.

After a good deal of thought and some slight experiment, I have preferred to rely almost *entirely* on my own memory for everything that follows. Otherwise, there is much difficulty (for my sort of 'scientific' mind) in avoiding getting bogged down in detail or over-cautious with trivia.

*   *   *

Bario is a long-house much like any other, looking down at it from the air at this moment, perched on the hillock where it has survived (off and on) for some centuries. Rather under a hundred people ordinarily live in it. The nearby long-house at Pa Main has over two hundred; that is big for a hill house. On the rivers, where transport gives mobility for people and harvests, long-houses may run to over one thousand; one house in the Batang Kayan runs over half a mile long.

Bario runs for two hundred and fifty feet, east and west. A track from the easterly verge of the plain winds through the network of irrigation ditches on to the bright green knoll, up a fourteen foot log, with notches cut into it to take the bare feet; through a three-foot hole, for the body to emerge, uncrouch itself, manoeuvre in the carrying basket and rub eyes smarting from the bright plain light; and already feeling the first touch of smoke which makes the inside's darkness white. If anyone were ever so rude—and the Bario people so impossibly negligent as to permit it—he could then pursue this track over the one hundred and more planks of the long-house verandah, out through the other little door at the upward west end, down the slightly sheltered notched log there (for knoll slopes upward already) and continue away into the mountains for a three week's trek to the nearest Government Station, bottled beer and taxation. The long-house *is* the path. In proper practice the newcomer sits down on a slightly raised platform at one side of the long verandah. Physically he relaxes; psychologically he tenses. Or a she even more so of both.

This verandah (for there is no better word for it) is thirty feet wide, running the whole length of the house. On the inner side is a continuous plank wall, reaching between eight and twelve feet high, where it tails up into a confusion of little platforms, cross beams, rice bins and fruit baskets, slung skins of honey-bear and clouded-leopard, piles of rush leaf for mat making, dried ears of maize for next season's planting, fish traps, spear hafts, and unintelligible agglomerations; all hung, strung, perched and crowded, as a false ceiling overhead. Above rises the true roof of delicately sewn palm thatch, the gable ridge sixty feet above the ground, darkly obscure until the gum-candles are lit at night. The roof thatch slopes evenly outwards and the eaves overlap a low wall which encloses the outer edge of the verandah, against the fierce winds, night cold and driving rain of these equatorial highlands.

The only piece of 'furniture', above floor level, is a little rough stool carved with a hornbill's head, the perch of Chief Lawai Bisarai, who is, however, more usually to be found lying down on one of the many red and white mats which are the true furniture—and which are made and traded from the Bawang two days to the east.

This verandah takes the ordinary traffic flow of Bario from a people exceedingly restless. Anyone can and must come in and sit down wherever he or she chooses along the verandah. A complete stranger will usually sit at the near end, immediately on the left as he comes up from the plain. (Few *strangers* will come the other way, where in a week's travel the only people are in a few small long-houses.) From this humble position—for the ends of the house are humble—unless something is known against him some one of the several people always about the house at any time will come and greet the new arrival in the neat Kelabit way. The words are chanted by the resident, repeated in a higher key by the visitor:

'You have come?'
'We have come.'
'You have come from?'
'We have come from Pa Trap.'
'You are many?'
'We are three.'
'All is well at home?'
'All is well.'
'Good—and all was well on the journey?'
'All was very well on the journey.'

After this essential exchange—and no more—tobacco will be brought out. Then the visitor is usually either asked through one of the four small doors in that long central wall, or moved up to the enlarged platform half way down the verandah—the upper-class focal part of the long-house where the Chief and his family reside.

Many other things happen on the verandah, and young men who have reached the age of energetic sexual enterprise officially sleep there. But there will be plenty of time to see more of all that. So go in with the invited traveller, crouch again through the low door of the long central wall, to emerge into strictly parallel but vastly more elaborate human arrangements.

For the dominant Kelabit arrangement is living in public, then planning inside. Nearly all other Borneans live either in separate houses or in long-houses with separate rooms all along the back-skin behind the verandah. There is not a lot of difference between living in a Kenyah or a Kayan long-house and living in Davenport Street, Bolton, or any other industrial housing. In Bario this is the big tangible difference. The back part of the house is as open as the front. There are, of course, separate and recognized family units. Each family group has its own fireplace, situated on this inner side of the long centre wall. On either side of this fireplace are plain mats—that is, beds. Around this fireplace the family (or such part of it as is around at the time) will eat in the early morning and again after dark. But even one fire may be and often is shared, for short or sometimes indefinite periods, by more than one 'family unit', each with its little allocated territory—observed by adults, more or less—around the shared hearth. And the stretch of ten feet or so away from the fireplace up to the back eaves is always open territory, the highway of all communal comers by night and day, adults and children, invited visitors, dogs, fowls, cats and house-cleansing cockroaches galore.

Each family is responsible for initially building and regularly upkeeping its section. But the house as a whole is a joint enterprise. Although all parts of it are individually owned, as a whole it is everybody's. There can be no thought of prohibiting anyone entering any part of it. And to a considerable extent personal property is mobile within it, also. Many things must not be taken and used without asking the person who originally acquired them. But some things, like axes (two) and other scarce iron tools, move about the house from end to end and fire to fire with almost anthropoid vitality. Others—like the single block of hard stone which acts as anvil and was brought as a joint effort from over the mountains into this plain of friable sandstone—stay wherever convenient. The visitor not very familiar with recent Bario deliveries would probably be unable to detect which children belonged to what families. Sometimes a child will not eat or sleep at his parent's fire for nights at a time. The kids roam the whole house; and—when they are old enough—on the plain around, with an independent but responsible mobility confounding close definition.

In this diffuse mood, a single sort of furniture dominates the interior landscape of Bario. All down the back wall of the house, in a continuous line, are the firm shouldered shapes of big, brown-glazed, eared jars. The older ones made in South China maybe over a thousand years ago; the not so old about Sawankalok in central Siam some six centuries back; the latest by Chinese immigrants, working in the old southern tradition on the coasts of Borneo within the past century. Around these jars centres a great deal of expressed Kelabit feeling. Out of their capacious depths flows a gratifying, sometimes seemingly continuous, stream of intoxicating beverage, the essential fluid of Bario's verbal dynamic.

*   *   *

Balio means petrification; and something more than that. Turning into stone is mixed up with the cold end of everything, although not verbally clarified or 'worked out' to that extent. Balio is the finish of a people, their domestic animals, belongings and buildings. It never happens to wild animals. It only happens to humans through their own 'fault'. It is the end of growing old-cold. It *could* happen suddenly; but the procedures causing petrification are hardly likely to occur except in a community which is losing its standards of proper Kelabitry anyway.

The principal causes of petrification are these:

Most often—ridiculing wild animals.

Often associated with this—harshness or lack of hospitality to orphan children looked after by grandmothers, who, in secret fury, put the rest of the long-house on the spot by presenting them with a wild animal in a ridiculous situation, so that they all laugh.

Very rarely (and rather uncertainly)—incest between father and daughter or some other extraordinary act; but as the petrification follows immediately upon the act, which in this kind of case will be secret, it is far from certain what is the cause, except in cases of animal ridicule, where there is usually at least one survivor (grandmother and orphan always survive, and by incestuous inference then renew the line).

The records of petrification are not preserved in song, but strictly in spoken stories; these are numerous. Stone stories tend to follow one pattern closely. Before hearing one of these curious stories—which always arouse interest whenever they are told for the benefit of children and strangers—a word on the process.

Petrification begins immediately after the act which triggers it. As this may have been performed away from the longhouse, during the day, the first warning most people get is a sudden drop in temperature. This is quickly followed by raging wind, hurling down freezing cold lumps of rain, hail, which is also Balio. Hail is extremely infrequent (snow unknown), even in this the highest settlement in Borneo. Its appearance justifies near-panic. Three things must be done at once. Everybody must get indoors. Every gong in the longhouse must be beaten as loudly as possible. And at least one of the two doors out from the verandah must be blocked with the most valuable, aristocratic and oldest dragon jar in the place.

The clamour of gongs, accompanied by wailing and crying, must continue as long as the hail. It is possible that this alone will prevent the long-house slowly turning into stone. But if that does happen all those who got safely in have now a second chance. For when once the hail has stopped and the petrification settled, everything is sealed in an enormous rock *except* that the venerable stone-ware jar cannot be conquered by stone. It is stronger than anything else anywhere. So when all other hope is lost and the people trapped, the owner of the jar shall smash it. Where it was fixed in the small square doorway, there is now a round hole as wide as the jar at its widest. Through this all may crawl out into the open and safety. Everything else they possess will have been lost. They cannot go back again; that night the hole will probably be blocked by fallen stone, anyway, leaving only a crack or mark to show future generations the way of the escape.

A long-house which is so lacking in high aristocracy that it has no sufficiently old jar has had it in any such event. Jar or not, more often the hail comes too quickly for any precautions, and everyone perishes. Those people and cattle who have got into or under the house are petrified where they cower, smaller stones dotted around.

With the hail comes a tearing wind, roaring thunder, searing lightning. After it, days and nights of continuous rain, which threatens to flood the whole land. Thunder always

241

causes some concern in consequence. The proper thing to do, when it starts to thunder, is to make small-scale thunder and echo noises yourself, crouched down, keeping still, hands over face. Which makes quite an inside din, in a storm.

Tradition tells of disastrous floods separate from petrification, but also caused by human acts. These have seldom been so serious; while the acts have always been corporate. Every Kelabit long-house has suffered at one time or another in its remembered history from an occasion where some sort of man-eating monster, tiger, or gigantic snake was slain and eaten. This beast lived in a hole, among rocks on the mountain side or in the bed of a river. From this hole it would emerge and stealthily gobble up first children and then grown-ups, gradually over months gnawing deep at the population. No one knew quite what to do, until some cunning young man thought of a trick to make it leave a trail—charcoal attached to live bait usually (the live bait always being recovered alive in the end). The monster was thus finally trailed and slain in its lair, usually with fire and smoke supported by spears. The killing of one of these monsters is commemorated on a big sloping boulder in the bed of the main Baram river, below the junction of the Libbun near Pa Dali. This dragon, incised in the stone, has the body of a bulging crocodile magnified, the head of a dog, eyes the size of big blue-and-white drinking bowls.

Flood troubles were not directly due to any killing of man-eaters. Unfortunately on several of these occasions, the delighted killers celebrated their vengeance by deigning to eat the corpse. The body, almost as long as the long-house, was cut up and shared out by fire-places. While the flesh bubbled in earthenware pots all along the house, the water boiling in the pots whispered this message:

'Beware, beware, you boil me here as your share. I have eaten you and you would now eat me. But if you do, you die. Beware, beware, beware.'

Only one hears this message; one family, poor people at the far end of the long-house (who have in some way been almost outcast). These heed the warning and escape swiftly into the hills or up river, before the feast begins and disaster falls upon the rest. Thus only the slaves are saved to start the social cycle again!

Somewhere in all these dreadful happenings, unnatural

behaviour towards living objects always comes in. One everynight style of petrification story goes:

'There was a great feast on in the long-house. But in all the fun and hospitality, no one remembered the old grandmother and her little orphan grandchild. They were left out of things. And indeed someone, drunkenly, made a joke about the old woman. The drunk gave her a piece of sugarcane, amusingly saying this was her share in the buffalo flesh and seeing of course that she had no teeth left strong enough to chew such stuff.

'Inside, the old woman went cold. So she told her little grandson to go down to the riverbank to find a nice, fat frog. He brought several back, she chose the largest. Then she took a tiny piece of cloth and wrapped it round its middle. She took two tiny bells and tied them round its feet; and another round its neck.

'She took a small brown jar, of the kind occasional borak [rice-wine] or strong arak are stored in. She dumped the frog into the empty jar, tied a piece of banana leaf over the top, as if there was a drink inside. She waited a while until the festivities had reached the height of frolic, levity, shouting, singing, dancing, arguing.

'Then, very humbly, she came forward into the centre and offered this tiny jar—where the long-house was oozing with the drinks of big ones—as her poor contribution to the feast. Her gesture was treated with some pleasure and laughter and all the rest. She gave it, as one can 'give' a jar to a particular person (who must presently reciprocate) to the man who had hurt her before.

'Soon he undid the leaftop, taking a gourd-spoon to ladle out the drink. Instead, out hopped the frog. The frog hopped across the floor looking ridiculous in its miniature skirt and bells. At the sight and sound everyone fell to laughing, cheering, all sorts of wisecracks.

'While this was going on, the woman quickly carried her grandchild out of the house, and away into the mountains.

'She was just in time. For the ridiculing of the frog came a tremendous sudden tempest. And before anyone could even think to place one old stone-jar in one of the housedoors the whole place was turned to stone.

'Not that it would have been much good anyway, since all jars were wet with borak, and that is no antidote.

'If you pass that way by the great rock you can still hear, deep inside, the sounds of hollow laughter (if you listen hard enough).'

So it is that stone, which now has little importance in the everyday essential routines of Bario life, in one way and another sandwiches that life between undertone fear and overtone ambition. At one side of this sandwich, petrification; at the other, those memorials for death which are the outstanding and distinctive characteristic of these people and the focus of most of the surplus energy, property and intelligence.

It could be that the rich plain set in the far folds of high mountain is somehow very slowly turning more and more into stone; or, that the human proportion here is under steady push from that direction. It is not simply coincidence that among the labyrinth of ideas which intertwine in intelligence, the two which have evidently long been of exceptional strength and intricacy, connect up a general theory of stone formed from human wrong-doing on the one side and a practice of carrying stones long distances, arranging or carving them for the respectable dead, on the other: the two lines of thought running in parallel zig-zag. These ideas intertwine with the general idea of nature and its relationship to human life, which is serious. Further, with the idea that in the past animals have been hostile to humanity on a big scale—and even in defeat have brought disaster when treated with contempt (cooked): treated properly, they also can stay memorialized in stone.

Aristocrats who have reached a reasonable age, and preserved—and by their industry maybe enlarged—their property, will ordinarily be memorialized either in stone or in those alternatives of a cut in the ground or river or through the forest across the crest of a mountain. The particular design followed depends on several things—including: the stated preference of the person directly concerned; the wishes of the family; the size of the feast to be given and the number of guests (equals potential labour force) to be expected; any particular needs of the community in that year; fashion, recent trends among other families of similar status; fashion, the urge to do the same but more; offsetting fashion, the desire to do something different; indications, conveyed to the experts through omens and dreams; weather at the time of the feast.

The scale of such enterprise can be measured crudely by the memorials which survive rain and flood, landslide, jungle growth and general decay. There must be nearly a thousand still recognizable within a day's walk of Bario. Nearly a hundred of these can still be associated with some definite persons. If everybody could afford to throw sufficiently large feasts, there would, clearly, be many more memorials. But to command the labour to carry and erect a monolith or climb the ranges to cut among the clouds, falls within the spending power of few. It is not that the labour has to be paid. It is simply that the feast must be big enough first to attract many guests and second to entertain them so well that they feel obligated—without feeling grateful—to do well by the host for one of the three or four days a big feast should last.

A big feast thrown by one aristocratic family as principal is likely to include and incorporate death rites due by less well-off families in the same long-house, simultaneously. These will contribute as they can. They will memorialize their dead more simply, that's all.

Bario does not bury its dead, unless for special reason. The proper procedure falls into two main parts. First, at death the deceased is laid out and keened over all through the night. This keening should continue without break until relatives from other long-houses have been summoned and see the cadaver. The sounds of grieving pour out across the plain. But they do not preclude a measure of festivity. Directly at death the family have asked friends of neighbouring fire-places to start preparing several big jars of borak in aid of the second phase.

The second phase coffins the body, which cannot well be kept out any longer in this climate. The principal mourner has commissioned others to prepare a coffin, the head and tail of which are carved (usually) with antlered figures, the whole shape something like a boat with a lid. When this is ready—worked in softwood, it can be done quickly—the dead one is put in, with appropriate additional uproar. The coffin is laid on the verandah; in front of the family fireplace on the other side of the centre dividing wall.

Here it will rest for at least a year. In the early stages the aroma is sufficiently conspicuous to the visitor. It has to be

'Mausoleum of Rajah Sinen's Family', from Carl Bock, *The Head-Hunters of Borneo: a Narrative of Travel up the Mahakam and down the Barito; also Journeyings in Sumatra*, Sampson, Low, Marston, Searle and Rivington, London, 1881.

taken as breathed by the residents, finding its natural level among the many mixed scents of the long-house, where cockroaches inside and pigs below act as effective scavengers, an unending chain of refuse disposal—with the advantage that the refuse also fattens the pigs for the death feasts.

To circumvent the smell so far as is possible, a long bamboo drains from the bottom of the coffin into the ground below the house. The coffin, if decently finished, fits tightly above. But when there are—for some reason, such as an epidemic—several dead in the long-house together, conditions do become pretty fierce for a while. Pun Maran notices it. However, to remove the dead from the house at this stage is not on. It would expose him or her to dereliction; and the living to haunting anxieties—perhaps to growing colder, even.

The third stage of the journey of the dead on earth: removal out of the long-house, away from everyday human activity. One of the functions of death feasts is to facilitate this spiritually tricky transfer, and to ensure that it has no spiritual repercussions of an undesirable kind.

There are a number of different ways of placing the remains at the end of the affair. By now the coffin on the verandah will contain nothing more than dry bones, beads and other personal ornament. These are removed and placed in a basket or jar, usually. If in a basket, to be placed in among rocks, or in stone vats and other arrangements.

With all due rites of spirit propitiation and proper lamentation, the bones are carried to the common place for final burial. It is not in the least compulsory to use this place, though. One Bario family keeps a private lean-to shelter, on the hillside nearer the long-house, for the bones of its departed kept in a special dragon jar (of Siamese origin). There are many variations. The common factor is a distinction between the whole body and then decomposition into skull and bones—the former inside the long-house, in the continued line beyond living; the latter away in the wild, with the living tie severed by extensive acts to reassure death (and deathliness).

These extensive acts are on such a scale that they form the hub of the non-rice year, a great outlet for social, family, and enemy relationships throughout the interior. Bario cannot have one of these feasts annually. No one could! This would put devastating strain upon the economy and energy of the community. All being well, anyway, someone need not die in the year—no one of importance, surely. And one occasion can do the deaths of five years (much more at a pinch). When someone of much importance has died, this is the time for others to attach themselves and carry out a combined operation, planned to be completed after the next or next below following rice harvest with the removed bones; provided, of course, that the harvests are good, everyone relaxed, abundant food and drink available.

Stone may be able to move by itself or arrive from the clouds. Humans only move it with efforts and purpose, some of the maximum effort of upland activity. It stands to Kelabit reason, therefore, that round these efforts centre the optimum intellectual organization of the people.

The system of stone and related memorials derives from property, class, occasion and inheritance. With people who do not write, inheritance is always liable to become complicated. In addition to the ordinary complications, Bario invites these:

— The wealthy have a great deal of property.
— Kinship and marriage rules are very considerably elastic.
— No kinship contract is entirely binding.
— All other contracts are variable and through inter- mediaries.
— The essence of upland character is to complicate, vary, and adumbrate.
— Property is only partly kept by its supposed owners and largely invested in loans to others, repayable with interest on demand.
— There is no marriage settlement of the *brian* type so widely used to adjust property between husband and wife sides of the family.
— There is no last will and testament—nor any other final and definitive statement on anything whatever at Bario.

The great death feasts are the Kelabit attempts periodically to clarify this situation, define heirs and inheritance, and cir- culate property (at interest) as widely as possible. No one can fully understand all the ramifications even of any one such activity, but every one thinks they know a lot about some. Each and every such occasion provides the opportunity for immediate or subsequent—sometimes generations sub- sequent—argument.

World Within. A Borneo Story, The Cresset Press/Hutchinson, London, 1959, pp. 3–4, 22–5, 114–21.

Recent Travels in Borneo

16
A Brief Visit to the Ot Danum Dayaks of West Borneo

MARIKA HANBURY-TENISON

In March 1973 Marika Hanbury-Tenison went with her hus-band Robin, the explorer and founder of Survival International, on a long and arduous journey through the Indonesian archipelago. They visited the islands of Java, Sumatra, Siberut, Kalimantan, Sulawesi, the Moluccas, and New Guinea. At that time Marika had only recently begun to accompany her husband on his expeditions, her first being a 12,000-mile journey to the interior of Brazil, in 1971. A Slice of Spice *records her second adventure. Robin Hanbury-Tenison had wished to familiarize himself with the situation of the tribal minorities of the Indonesian archipelago, follow-ing his work on behalf of the South American Indians.*

Marika and Robin travelled to the interior of West Kalimantan. They flew to Nangapinoh and journeyed up the Melawi River in the company of a local guide, Amri. This extract, Chapter 7 of the book, describes her brief visit to the Ot Danum Dayaks, and provides insights into some of the changes taking place in longhouse life and Dayak culture.

Apart from her well-written, humorous, and thoroughly entertaining travel books, Marika Hanbury-Tenison was prob-ably best known for her cookery books, and as the Cookery Editor of The Sunday Telegraph. *Sadly, she died in 1982 when she was still only forty-three years old. My wife and I happened to meet Marika and Robin Hanbury-Tenison during their visit to West Kalimantan, and we shall always remember Marika's generosity of spirit and her zest for life.*

THIS time, I managed to negotiate, without disaster, the bridge—a long, narrow trunk with notches as footholds—and a slippery stile across a wall of sharp bamboos and the final high ladder to the centre of the house. An old man came to meet us; Amri explained who we were, and we were led down the central passage dimly lit by small kerosene lamps to the very end of the longhouse. Our footsteps echoed on the floor of bamboo poles, faces peered at us from doorways along the sides, and the shapes of large wooden storage chests and heavy troughs for pounding rice were shadowy in the half-light.

The old man indicated we should sit in the corner on finely woven mats of thin rattan, shiny with use. Too late, we realized we should have removed our shoes, as Amri and the boatmen had done, before entering the house. Now a crowd of a score of more silent men, wearing sarongs, their faces expressionless, watched as we struggled to undo laces clogged with mud.

A brighter lamp was brought out from one of the rooms and the silence continued as we sat cross-legged, the circle of men around us, and offered round packets of tobacco and Kretek cigarettes. They regarded the tobacco with suspicion, sniffing it and placing it untouched on the floor, but the cigarettes seemed to be more successful and soon everyone was smoking hard, thickening the air with a grey haze as they continued to stare.

What were they thinking? The Dayaks had been headhunters; some still were. Were these people upset by our invasion: strangers, uninvited? Were they friendly or hostile? It was impossible to tell from their faces and the glint of steel blades from long, pointed knives began to unnerve me.

Then, at last, the old man began to talk to Amri and the atmosphere relaxed slightly. More cigarettes were offered and accepted, a few of the men began shredding the tobacco in their fingers and rolling it into the thin paper we also had with us.

'They have little food here,' Amri reported apologetically.

'Perhaps I could give one of your sardine tins to the women to cook.'

So far, we had not really seen any women. They remained hidden in their houses along the passage, peering out as we followed Amri down to the river to wash and stepped over some of the fallen idols lying on the ground. They shone silver-white in the moonlight and the squarely carved faces and bodies looked strangely real and human. Around us the nightly chorus of frogs and cicadas was in full swing, and above us flying foxes by the dozen were silhouetted against the clear, starlit sky.

We were in Indonesia to find out how the isolated tribes of that country were situated in the twentieth century, but it was not easy to get close enough to these people to understand them fully, and once more time was at a premium. Very few of the Dayaks spoke more than a few words of Indonesian and, although Robin's command of the language was improving daily, local accents complicated the Malay-based language. Their controlled features, though full of dignity and politeness, were almost impossible to penetrate.

The Bataks in Sumatra had been almost aggressively open; the Mentawai had been naively and enchantingly uninhibited; these people were extremely isolated and yet, because they had the communication of the river Melawi on their doorstep and had, for centuries, traded with the Chinese, they had a highly developed cultural tradition. They displayed the minimum of emotion through their features and, at first, talking through the interpretation of Amri, we felt how totally foreign and alien we were in their midst.

Although it was about nine o'clock, work for the day was only just finishing and a stream of men, their sarongs rolled high around their waists, the rest of their bodies bare, came into the house carrying sacks of rice and palm nuts. They unrolled the lengths of cotton cloth, pulled them out to the full width, crossed the slack over and neatly tied them tightly around their waists again as they came to sit with the other men. Women now appeared from their houses, dressed in their best, in colourful sarongs and tightly-fitting jackets, hair tightly scraped back from their high foreheads. Again, they sat by themselves, brought out brass trays bearing an assortment of small pots and busied themselves by preparing betel nut. This was accomplished by crushing the nut (a round, hard, brown knob the size of a walnut) and placing it on a leaf with a pinch of lime, then rolling up the leaf and popping the whole thing into their mouths, chewing methodically and occasionally spitting out the juices through the gaps in the bamboo floor. Their gums were red from the nut, their cheeks distorted as they kept the package in one side. Some of the women would have been beautiful but their otherwise fine features and smooth nut-brown skin were spoilt by teeth discoloured and rotting from lime.

One of the women brought out a tray of glasses filled with sweet acrid coffee, thick with grounds, and sank gracefully to her knees as she offered it first to Robin and then to me. When the coffee was finished, she disappeared again and this time returned with a brass tray on which there were a dozen glasses and a large earthenware jar. From it, she poured a pale milky liquid, offered it to us and then filled the remaining glasses, passing them round amongst the men.

'Palm wine,' said Amri, grinning, 'very strong, very good.' It had a sour, bitter taste but was cool and not unpleasant. When that jar was empty, someone fetched another and the pouring began again. The women did not drink, concentrating on their betel nut.

Men were still coming up into the house with sacks and baskets on their backs, many of them tired-looking and dirty. By now, there were about a hundred and fifty people sitting up our end of the long, long room which seemed to stretch into infinity in the darkness. The air was thick with smoke and heavy with the smell of wine, the pigs under the house and betel nut.

A set of gongs was carefully carried onto the mats and some children, a few of them now dressed in cheap, skimpy cotton dresses instead of their sarongs, brought out a large pole with three crossbars and hung with chains of coloured paper. Spiders and other insects, disturbed by the activity, rustled in the palm-thatched walls behind our backs and ran scuttling across the rattan mats.

The people were relaxed now, talking softly amongst themselves, often turning to smile shyly at us. Some of the men rubbed muscles which ached after working in their rice fields, most of which were up to three days' walk from the village. Many were pitifully thin; there were notably few old people and only one man with grey hair, but the young men looked strong and the women, although their faces were often drawn, had a grace that I envied.

One of the young men produced a strange instrument made from bamboos and small gourds, fashioned rather like an organ. At the bottom was a gourd with a curved stem like a maraschino pipe through which he began to suck, his cheeks drawn in, his body unmoving in total concentration. There was silence again and strange music filled the room and echoed down the long passage. The sounds were soft and haunting, a low reedy bagpipe, sighing with notes that lingered and undulated. After a time, he was joined by another man, this time with a crudely-carved, elongated banjo, playing softly in accompaniment. Then the gongs joined in, tinkling and light, and a group of young girls danced around the colourful pole, swaying in time with the music, their hands twisting and curving.

Now the stage was taken by a man wearing only a sarong, richly coloured in greens and reds, tied between his legs to make a loincloth. On his head was a large and complicated turban. At his side was a long dagger sheathed in an ornately carved wooden case.

'This is a war dance for when the Dayaks are attacked by the Punan hill tribes,' Amri whispered, and we watched fascinated as, very slowly, the dancer began to perform the complicated steps of an ancient ritual, pivoting on the sides of his feet, arms twisting smoothly, crouching to the ground with a hand shielding his eyes as he looked for the Punans. Slowly the dance built up to a crescendo, faster and faster as he leapt into the air, legs outstretched, with a great cry of: 'The Punans are coming.' He drew his sword, which glittered as the light fell on a razor edge, swinging it round with more cries as he killed one after another of the enemy and then sprung once more into the air with a leap of staggering height as they were at last defeated.

Again there was no clapping and the crowd settled themselves more comfortably, accepting more cigarettes and sipping at the palm wine. Many of the men were now chewing betel nut too, taking time over the ritualistic preparations; some were getting glassy-eyed and the air was so thick with smoke it stung my eyes.

There was a sway of bodies as another dancer moved through the crowd to the centre.

A woman's sarong, a tight-fitting white lace jacket and a beautifully embroidered scarf tied gracefully around the head, hiding both hair and face, did not totally disguise the fact that this was a man, although his dancing was totally feminine. It was a woman's dance with a woman's motions and an over-riding aura of sex. Hands curved and fluttered with female elegance, hips gyrated in a sexual flowing movement in time with the singing wail of the pipes, gongs and the string instrument. He pivoted and swayed, sinking and rising to the ground with pelvis forward and his back arched, the scarf never once moving to expose his face and the jacket padded in some way so that even the breasts were those of a woman.

It was a transvestite dance performed as part of a culture and tradition, not for amusement. No-one laughed as the crowd sat motionless, watching.

After more children danced, their childish movements showing signs of the grace that was to come, a second man gave another beautiful performance. He was straight from the fields, hands calloused, face still streaked with dirt, and he wore ragged torn trousers that were caked with mud. I found it amazing that a man who had laboured all day could then transform himself into a thing of such exquisiteness. Again the dance was effeminate; many of the movements reminded me of Thai dancing I had seen in London as his feet remained on one spot and he pivoted slowly round, his body curving and arching and his arms waving slowly. The incredible movements of his hands and fingers must have taken years of practice to perfect.

The dancing ended, but the party seemed to be continuing and the palm wine was still being passed round, although it was now into the early hours of the morning. Both of us were yawning, I was aching from sitting on the floor in the same position for what seemed like hours and hours, and my head was spinning.

Amri, our faithful courier whom we were getting to like more and more, came to our rescue. 'You want to sleep?' he asked, and, looking round the crowded open area of the longhouse, decided we would be better in one of the houses. We left our luggage where it was. We were slightly nervous about it but Amri assured us it would not be touched.

In an eight-foot-square room at the back, thin mats were unrolled and straw pillows, covered with coarse, heavily embroidered linen cases, were laid on the ground. We had nothing to cover us as our knapsack and the thin blankets inside it were soaked through from the rains.

Although it was swept and clean, there was a smell of dust in the room, and we could hear the noise of the Dayaks, talking, spitting out their betel nuts and occasionally singing. A dim light filtered in through the plaited walls and I could hear the scurrying of small animals moving along the edges of the room and in the roof above.

As soon as I lay down, I felt sick. Desperately, violently sick, with gorge rising into my mouth and a sour taste as it filled my mouth. It might have been the palm wine, but I had not had much to drink. I was more certain it had been the few chews I had had of another revolting durian that I had to accept out of politeness earlier in the evening.

I began to panic. I could not go out into that crowd in my nightdress and risk being sick in front of them, and although I shone my torch along the boards there was no crack large enough to be sick through. I sat up, I walked around the tiny room, and I stuffed a handkerchief into my mouth, watched by Robin who was powerless to do anything except hold my hand.

Hour after hour, it went on. Wave after wave of nausea swept over me and each time I fought back one I was sure I would not be able to control the next. Outside, the noise of the Dayaks talking and singing continued.

Sleep finally conquered. At last I was able to lie down again and we huddled together on our mats like two puppies as, by then, I was cold and shivering.

The revelling of the night before had obviously had its effect, as we were up before anyone else seemed to be moving the next morning. We took a walk past two smaller long-houses, also built on high stilts—protection from attack from hostile tribes—across a fragile bridge of bamboo, across another bamboo stockade and into the forest surrounding the village, disturbing an occasional large, coarse-haired pig not far removed from its wild cousin, the boar, long-nosed, small-eyed and viciously tusked. From the jungle we could hear the sound of monkeys chattering to each other from tree to tree.

In the daylight, the carved idols looked less spooky. They

were dotted in amongst coconut trees, some obviously very old indeed, primitive facsimiles of men wearing loincloths with their hair in a sort of top knot. More modern carvings depicted dead ancestors wearing crudely fashioned shorts and one sported what looked extraordinarily like a bowler hat and carried what could only have been a pair of binoculars in his hands. Some of the poles had offerings of earthenware pots and gongs tied to them, green with age. A few had been roughly painted in a now-faded royal blue and turquoise.

In one of the smaller houses we had a breakfast of rice and a little thin soup, drinking boiled water that was grey with mud, and then joined the chief who sat cross-legged on the floor with his wife and two older women around a betel-nut tray. We gave them some tea, coffee and soap which seemed little in return for their hospitality and the dancing we had seen the night before, and they accepted the poor presents with dignity. As they accompanied us down to the sampan, they asked us to return again, or to stay longer.

The river was even higher that morning and seemed to be moving at an incredible rate, swirling around debris brought down from the mountains—trees eighty feet long, pulled up by their roots. Soon the sun blazed down in full force and this time we had no shade as frequent, lashing rain had ruined our plaited canopy. I could feel myself burning up but at least our clothes and kit dried out.

Before long, the river narrowed and we battled through tumbling rapids past banks that were masked with thick jungle, bamboos and trees overhanging the water that had strange, fleshy plants growing on their branches and were festooned with vines. On the opposite bank, a strange, large bird, the size of a turkey, flopped clumsily amongst some bushes, half hopping and half flying, with a large, russet-brown wing span and a fan spread tail. Once we saw a whole tree, stripped of its leaves, covered with monkeys as many as twenty to a branch, swinging from their tails and scratching themselves. It was a fantastic sight and I scrambled carelessly to a crouching position with my camera poised, the sudden movement almost upsetting the shallow craft.

Around mid-day, we chugged round one of the twisting corners and arrived at a surprisingly modern-looking collection of brick-built houses, glaring white in the sunshine, each with a little garden in front of it. This, Amri told us, was the

Camat's [sub-district official] office of the area and once again we must report and show our passports and our passes.

The office was surprisingly large and one wondered what on earth the eight Indonesians sitting in it could possibly have to do. Their two large typewriters were dusty and none of the faded notices pinned to the wall were dated later than 1971. By now, we were getting far enough inland to be within reach of the Sarawak border where, since confrontation, there had been constant fighting, so perhaps this post of Ambalou was there in case of trouble. All the men wore pale khaki uniform but it was impossible to know whether they were police, soldiers, or merely members of the Camat's staff. They laboriously copied down every page of our passports and gave us coffee, strong, delicious and very welcome.

By late afternoon after frequent heavy rainstorms we finally reached the village of Sabun which, as it turned out, was the end of the trail. On the few vague maps Robin had seen, it had looked as though we might have been able to go another ten or fifteen miles up the river, but this was obviously going to be impossible by boat. Above a collection of six long houses on the banks, the river rose in a series of rapids so steep they were more like waterfalls with white water tumbling over a high mass of black rocks. Beyond us the river swerved sharply around a great fountain of giant bamboos and disappeared out of view.

The village had—to my delight—no raft or bath house here and no perilous log bridge. We tied up against the bank and followed two young women, their sarongs, soaking wet and clinging to their bodies, outlining shortish, strong figures. On their heads they carried large circular gourds heavy with water.

The houses were smaller than the one the night before but in much better repair, and Sabun had an air of prosperity about it, although here there were virtually no signs of Western or any missionary influences. The Dayaks too looked stronger and more healthy. They were much shorter and more stocky than the people we had met lower down the river and, in many ways, far more open and less reserved in their attitude towards us.

In the centre of the village was a strange shrine, a wooden platform on high stilts covered with a thatched roof that was obviously still very much in use. At each end a primitive carving of a man and a woman faced towards the village below them; in the centre was a grotesquely grinning devil face. On the ledge, offerings of plates with small mounds of rice were surrounded by dried pig's trotters, bunches of herbs and the skulls of wild boar, and what I supposed were monkeys. A woman passing it—her head covered with an incredibly wide-brimmed hat patterned with strange oriental designs—carried charms in her hands, more tinkling pigs' trotters and a string of strangely carved objects designed to bring good luck, as she went off to fish.

In the largest house we met the chief, an elderly grey-haired man with a high forehead and deep widow's peak, who invited us to sit and offered us betel nut from a highly ornate brass tray with the ingredients in equally intricately decorated small brass pots. Both Robin and I refused; the idea of chewing raw lime and the red-lipped results were too much for either of us, although we both tried it later.

Children played around the feet of their parents, quiet and well-behaved. The man sat cross-legged, in that yoga style I had not yet managed to perfect, with legs entwined and bare feet flat side up, and the women squatted on firmly planted feet, their bottoms almost touching the floor, backs ramrod straight, breasts exposed above their sarongs. The house was simple, unfurnished except for exceptionally fine mats on the floor and some knives hanging on the walls beside wide cartwheel hats, decorated with tiny pearl buttons, painted red and with tassels of human hair hanging in a pigtail tightly tied with coloured ribbons coming from the coned crowns.

We were sitting in the communal area outside the door of someone's house, and each time a member of the family or one of the children wanted to enter through the half-closed door, they had to push their way through about fifty people sitting on the ground outside. They did this with the curious deferential politeness we saw throughout the interior of Indonesia, half-bending to the ground as they passed and holding an arm out straight in front of them as they moved. This we learnt was an ancient tradition which had grown up to show they were not moving in anger; the crouch was an apologetic gesture and the outstretched arm was to show they were not armed. Even the smallest children performed these motions every time they went to or from the door.

One of the older men went to sit by the main door to the

longhouse and banged out a rhythmic pattern on an ancient brass gong.

'He calls the people to the chief,' Amri, who had changed from his uniform to the traditional sarong, explained, and within seconds the Dayaks began arriving, more and more; men, women and children squeezed into the small longhouse until there must have been nearly a hundred of them sitting almost shoulder to shoulder on the floor. Brass betel trays were passed between small groups and everyone was busy chewing, rolling and crushing the betel or spitting out saliva between the floor boards. A fascinated row of chickens and cockerels perched on the two window frames and blinked through red-rimmed eyes into the room. Our Chinese boatmen stayed very much in the background.

Though similar to the music we had heard with other Dayak tribes, the entertainment that evening had a purer, simpler quality to it and there was no dancing. Men played alone or in harmony on strange bamboo pipes and stringed instruments: soft, haunting notes that floated through the air, echoing, slightly, with a gentle, sighing sound.

It was a pleasant, soothing way to spend an evening but, much as I had enjoyed the music, I was glad when the party seemed to break up at about ten-thirty. The instruments were put away and, although no-one moved, that was obviously the end of the entertainment.

'I think they are waiting for us to do something,' I whispered to Robin.

'You may be right. Let's go and wash.'

We walked through the crowd, pushing out our right arms politely and moving with a half crouch, clutching our washing things in the other hand. It felt a bit silly, but 'when in Rome . . .'.

When we returned, the room was empty except for Amri and his boys already stretched out in a row at one end of the room. Beside them was a paraffin lamp, turned down low and surrounded by thousands of small insects and a few enormous moths. At the other end of the room, our luggage was neatly stacked against a wall and two of the most beautiful mats we had yet seen were laid out on the floor. The rattan was so fine it was almost impossible to see the weave; both mats were shiny with use and patterned with intricate black lines.

From the chief of this tribe Robin had learnt earlier in the evening that, even if we carried the sampan above the rapids, the river was virtually impassable even in this high water and, in any case, the villages beyond were few and far between and difficult to locate. He also told us of a sensational waterfall, about three or four days' walk away. Amri, who was helping to translate, said he had also heard rumours about this fall but, as far as he knew, no-one outside the Dayaks living in the area had ever seen it.

'I believe it is higher than Niagara,' he had said, and I could see Robin's eyes gleam. To be the first outsiders to reach such a waterfall would be an achievement indeed. The instincts of an explorer did battle with his instincts of trying to keep to our tight schedule. Was the fall really only four days' walk away at the outside? Could it be done in less? How tough would the walking be and could I keep up? Could we get a guide? All these were questions which had to be taken account of. If we went we would never be back in Nangapinoh in time to get our lift back to Pontianak and it might be days, or even weeks, before we could get the plane to come in again for us. Reluctantly, he decided we ought to return. We had already taken an extra day to get this far and, if we had trouble with the boat going back, we might not return in time.

It gets light early in the tropics and it seemed no time at all between falling asleep and being woken by the cocks crowing outside. After we had had some coffee, Amri asked me to go and look at a sick girl lying in one of the little houses built off the main longhouse. I took our medicine chest and followed him into a small room where the personal things of the family were neatly stacked around the walls beside large porcelain jars, used for storing wine, oil or the spirits of the gods; jars which had been brought up the river by the first Chinese explorers centuries ago and which, at Sotheby's or Christie's, would have fetched a small fortune.

The girl was obviously very ill. She lay in the centre of the room, her smooth, walnut-brown skin tinged with a greenish-yellow hue, her forehead damp with beads of fever sweat and her hands chill and clammy. She shivered constantly and I managed to find out that she had been very sick. She obviously had a high temperature.

Amri had not followed me into the room so I went out to confer with him.

'She should go to hospital,' I said.

'Yes, they will take her to hospital but it will take them three days. Can you give her something to make her better for the journey?'

What should I give her? We had all sorts of antibiotics and pain-killers but I did not know if the girl had malaria or polio. The people standing round her shook their heads when I said malaria, so I supposed they knew better than I would. Just in case, I gave her mother a supply of malaria pills, some codis and some stomach pills to stop her being sick. I told her carefully what to do with them; none of the medications could do her any harm and I just hoped they would help to make the journey more comfortable.

We went down to the river with her family to see them off. Careful preparations had been made for the journey. The shallow canoe was covered with an arch of freshly plaited green palms; gourds of water and wine were packed carefully in the back with a sack of rice and two live chickens. Two men laid the sick girl carefully on a straw mat, laid over more palm leaves and wrapped some freshly picked banana leaves around her head. Her father and mother picked up their paddles and joined her. A small child sat motionless in the centre of the small craft beside the girl's legs as the rest of the village stood on the bank to watch them go.

We re-packed and left soon after, sad to leave so soon this village in its idyllic setting and the people who lived there.

'Some day,' Robin said as the boat drew away from the bank and we waved to all the people who lined it to see us off, 'some day, we will come back here and find that waterfall.' It was a nice thought but both of us knew if we left it too long the village of Sabun and its surroundings would be sadly changed. The prospects of wealth from lumber and minerals were beginning to boom in Kalimantan. Already, much of the forest and jungle further to the west had been raped and the ground left barren; there had been a French team of mineralogists in Nangapinoh before we arrived and, if deposits were found, it would be people like these Dayaks up the Melawi who would suffer. With their land destroyed, the cloak of poverty would quickly encompass them and almost certainly little, if any, of the profits of this pillage would go into their pockets. Perhaps institutions like Survival International and others like it around the world could help to protect the rights and the lives from such pillage.

We had not got far before we met trouble. The rapids which had nearly swamped us the day before were even rougher now that the level of the water had fallen. We drew into the bank above them while Amri considered the situation.

'We will leave the baggage in,' he told the men. 'Mr Tenison and the lady will walk on the bank and we will pull the sampan through.' Robin was not terribly keen on behaving like a white Rajah—he liked to be 'where the action was'—but Amri was in charge and we obediently climbed out of the boat onto the bank.

Although the bank was heavily covered with undergrowth and our way often blocked by thick vines so that we had to wade through the water around the edge, we made better progress than the four men dragging the boat over the rocks. Three times it tipped sideways and I thought it would overturn, but somehow they kept it steady against the current and the rushing river and an hour later they were out the other side and into smooth water again. We slid across sandy boulders to join them, surrounded by a cloud of white and yellow butterflies, many at least six inches across from the tip of one wing to the edge of the other. They danced round our heads and settled on our shoulders as we walked and then flitted off as we stepped back into the boat.

We shot the next two rapids, careering madly through the water with waves breaking over the side of the sampan, swinging and bobbing with what seemed to me virtually no control at all. It seemed almost miraculous that nothing disastrous happened to either the boat or its occupants.

At Ambalou, we stopped to reload the petrol drum which seemed suspiciously lighter than when we had left it there. Robin swore a bit but there was nothing we could do about it, and Amri's usually fluent understanding of his Indonesian suddenly went to pieces. While we were standing on the raft jetty arguing the point in a friendly fashion, the boatmen disappeared into the jungle and came back with armfuls of small fruit with smooth, yellowish-brown skins. They heaped them into our laps with broad grins and Robin's sense of humour triumphed over anger at being gypped over the petrol.

'Oh well,' he said, grinning too, 'they always say that fair exchange is no robbery. These had better be good, these fruits.' Luckily, they were.

Once again, we had to check in at the Camat's office at the little whitewashed settlement below the village of Ambalou and this time we got VIP treatment. The Camat was wearing his best uniform, looking rather hot in a suit that had obviously been made some years before, and breathing heavily under the weight of a dozen rusty medals. His second-in-command was wearing his best false teeth, of shining white porcelain, which made his jaw jut and his ears stick out. They did not quite meet in the front but the back clashed with a terrible clatter.

We were invited into the Camat's house, tatty and cheap-looking compared to the longhouse at Sabun, with old pages from a prudish girlie calendar pinned to the wall beside a faded print of the President and an eight-inch-long, heavily armoured beetle, which one of his children had tied by a long piece of cotton to a shelf on the wall, buzzing sadly in the background, occasionally beating its wings with a whirring sound.

Lunch, by Dayak standards, was a splendiferous affair: huge bowls of rice, a dish of stewed chicken, hard-boiled duck's eggs and some peppery sauce. At the end of it, the Camat leaned back contentedly, making every medal pinned to his chest bounce up and down with the vibrations of a mammoth belch. The finale of the meal was glasses of a white liquid which I thought must be palm wine. No such luck. The Camat and his minions were Indonesians (probably from Java or Sumatra), belonging to the Muslim religion and therefore teetotallers. The liquid turned out to be our first taste of a speciality we were to be offered many times again on our journey—lukewarm, diluted, tinned milk, heavily sweetened and thickly sickening. I found it unbearably cloying and almost impossible to swallow. Robin, on the other hand, who has always retained a nursery palate, loved it. Often, he had to drink mine as well as his and we got very nimble about switching glasses so as not to cause offence to people to whom tinned milk was as precious as gold dust.

After lying motionless in the blazing sun for an hour, the boat was burning to the touch when we set off again. The petrol drum was so hot, it was positively dangerous, and it was difficult to get comfortable under the palm canopy without touching something red-hot. Sweat trickled along our bodies and a foot that Robin left carelessly dangling outside the shade soon turned an ugly red, but at least the motor kept going and we made good progress now that we were no longer fighting the current.

Nangapinoh—reached the next afternoon after a night in another village, where conditions were bad and the people near starvation, and then a long, hot day on the water— seemed like a return to another world. The feeling was intensified when, after checking in with the Camat, we hurriedly changed and walked to the outskirts of the town to take 'tea' with the American Missionary, Mr Van Patten, and his wife—an invitation accepted before we had left the town.

Their house was, staggeringly, straight out of *American Homes and Gardens* with a 'country flavour'. I felt as out of place in it as a prostitute at a deb dance, sitting on my hands to stop myself smoking, on the edge of a chair upholstered in pale linen. The room was coloured with pastel hues and contained cotton tablecloths with frilly skirts, 'snaps' of smiling children and little things from 'way back home'. I shivered slightly in air that was stirred by a fan and carefully screened from mosquitoes. Mrs Van Patten looked cool and ladylike in a knee-length cotton 'frock'. I felt hot and sticky and longed for a long, cold beer. My feet, sticking through open sandals, had a decaying, greyish tinge to them although I'd spent ages scrubbing them in the primitive bathroom of the Camat's office, using a piece of coconut husk as a nail brush. I hid broken, mud-engrained nails under the table as Van Patten said grace, calling on the Lord to bless 'our visitors, the Tenisons, who eat with us today' and asking Him to watch over our journey. 'Amen to that,' I said under my breath.

The Van Pattens had been in Indonesia for twenty-eight years and were able to fill us in on a lot of background history of Kalimantan. During the conflict between the government and Communism in 1965, they had seen many graves being dug for anti-Communists, had seen two prepared for their own bodies, and then witnessed the massacre of scores of Chinese as the revolution was crushed.

'Many Chinese are still being hunted down as suspected Communists,' Van Patten said, as we supped off a good

American meal of eggs scrambled with tinned meat, toast with cranberry jam and a salad of grated raw carrots and raisins. 'They hide in the jungle with remote groups and there have been many unfortunate cases of Dayaks being murdered to prevent them disclosing their whereabouts. Recently, the Dayaks rose up against the Chinese because of this and there was a lot of bloodshed. But of course it's often almost impossible to tell who is Red and who isn't in this kind of situation—everyone suspects everyone else.'

He told us about a young Chinese photographer who had been arrested for Communist activities the week before. He was the official Nangapinoh photographer and had gone on a trip with the MAF plane to take pictures of the town from the air. Soon after, some officials took him away on the charge that he had been at a party with some identified Communists some years before.

At that moment, the man in question walked in. He looked as though he had been a prisoner in a concentration camp, thin to the point of emaciation and exceptionally pale. Although he was obviously nervous about saying too much in front of us, he did talk a little about his experiences in prison.

'They didn't beat me up but they did interrogate me for over ten hours and they removed my film—all my films. I was lucky. It wasn't too uncomfortable and I could pay to be released.'

'How much?' we asked, but he shook his head and smiled thinly. That was something he was not prepared to divulge. I thought about the photographs we had taken of Nangapinoh and sincerely hoped no-one was going to think we were potential Communists.

That evening, we had one of the nicest experiences we were to have on the whole expedition, one of those moments when one is touched to the point of tears.

Our courier, Amri, who had disappeared immediately we reached the town, appeared again at the Camat's house and sat down with us on one of the dusty plastic-ribbed chairs still standing in a straight row down the central room.

'How was everything at your home?' I asked him. Amri smiled broadly.

'Everything was very well,' he said. 'My wife had a son. A big boy, very strong and our first man child.'

We were astonished. Amri had never said anything about his wife expecting a baby while we were away, although, now we thought about it, we realized he had been in a bit of a hurry to get home. He cleared his throat a few times, rose to his feet, clasped his hands in front of him and began speaking very fast, in English, as though repeating a well-learned lesson:

'I should like, with your permission, to call my son Tenison Amri Bachtiar, and for you, Mr Tenison, to be his godfather.' He hesitated and, obviously feeling I might be hurt at being left out, added, 'The Lady Tenison too, of course, would be very honourable.'

I did not know about being 'honourable', but both Robin and I were terribly touched. We gave Amri a copy of Robin's book about the Indians of Brazil, *A Question of Survival,* and promised to keep in touch with him and little Tenison. I would love to have seen the baby but Amri, regretfully, said his wife was not yet 'arranged' for visitors.

Just then we were joined by the Camat, looking neat and official in a sand-coloured uniform. He too obviously had something important to say because once more there was again a considerable amount of throat-clearing and some hastily whispered words in Indonesian to Amri, who went pink about the ears. Amri was chosen as the one to speak.

The people in Kalimantan, Amri said, were having too many babies. We had produced only two children in fifteen years of marriage. Did we take the tablets they had heard of which prevented babies? Did we know where they could get these tablets? Was it true that abortion was allowed by our religion? Both men kept their eyes fixed on the overflowing ashtrays on the shabby tables in front of us as the questions were asked and we tried to answer them.

These people were living far into the interior of an island which was itself far from the governing capital of Jakarta, yet they were highly aware, and desperately keen to do some-thing about a problem which the Western world has really only recently opened its eyes to. Their attitude was refresh-ing.

The conversation lasted for about an hour and then Amri left to get back to his family. We were sad at parting com-pany. At the speed we were moving, friendships tended to be fleeting things and in Amri we have found someone we considered very special. Apart from anything else, we had him to thank for the tremendous success of our trip and the ease with which we had been able to visit the Dayaks and their longhouses.

Van Patten took us out to the airstrip next morning and waited with us to collect the post from Pontianak.

Two Indonesians cleared the runway of cows and picked up their droppings to spread on the missionary garden, and after an hour's hot wait, with flies and mosquitoes buzzing annoyingly around our faces, the MAF pilot, George Boggs, arrived out of the sky. Within five minutes we were airborne with the wooden town of Nangapinoh, the houseboats (bandungs) and the small dagans (motor canoes) only a colourful blur in the distance.

A Slice of Spice. Travels to the Indonesian Islands, Hutchinson, London, 1974, pp. 83–102.

17
Up the Baleh with Kenyahs and Ibans

REDMOND O'HANLON

Redmond O'Hanlon's Into the Heart of Borneo *is one of the outstanding travel books of recent years. Witty and informed, it describes a journey up the Rejang and Baleh Rivers in Sarawak and the ascent of Batu Tiban in the borderlands of Kalimantan and Sarawak. O'Hanlon also travelled further up the Rejang River to the small market town of Belaga and beyond to the Baluy River. He went in the company of the Oxford poet, James Fenton, and three Iban guides, Leon, Dana, and Inghai. The perceptive, skilful, and humorous narrative is, in some respects, reminiscent of Andro Linklater's* Wild People *(see Passage 18), but O'Hanlon's story is also liberally sprinkled with references to various scholarly texts on Borneo. This is perhaps to be expected. O'Hanlon had obtained his Master of Philosophy degree in nineteenth-century English studies at Oxford, and he completed his doctoral thesis in 1977 on 'Changing Scientific Concepts of*

267

Nature in the English Novel, 1850–1920'. He had also published on Charles Darwin and Joseph Conrad, and became a Fellow of the Royal Geographical Society in 1984.

O'Hanlon had obviously done his homework on Borneo before embarking on his journey, ostensibly to find the Bornean rhinoceros. He reads on the way, taking with him such texts as Bertram Smythie's The Birds of Borneo *and Lord Medway's* Mammals of Borneo. *James Fenton prefers to read* Les Misérables.

This extract, which comprises Chapter 9 of the book, records the stay in a Kenyah longhouse in the Upper Baleh. There was obviously a relaxed familiarity in O'Hanlon's relations with his hosts and his Iban travelling companions. We also read in this extract of two other important elements of Borneo cultures: penis pins and bird omenism.

I awoke very suddenly, well before dawn. There were no cocks crowing; the pigs were snoring under the floor; the dogs were silent. Even the geckoes were asleep. But there was a light coming towards me, a taper-light. Someone was approaching from the very back of the room by the kitchen. It was a small, rustly, floaty, pinkish, graceful sort of figure. Leon lay a foot to my right, asleep. She knelt silently down and tugged his foot. Leon stirred.

'Shussh,' said the young girl, pulling him to his feet.

Leon muttered something in Iban.

'Shussh,' she said, and led him behind a partition on my left.

There were subdued giggles, murmurings, rustlings, kisses, squeaks. And then, it seemed, three hundred yards of longhouse began to shake. Leon, single-handed, with only a very little tender help, I thought, will have the whole lot down like a timberyard. The cross-beams rubbed back and forth on their supports. The joists strained at their rattan loops. The piles, perhaps, deep down below, thrust in and out of the earth. And Leon's watch spoke his triumph into the night: *Beeeep-beeeep-beeeep-beeeep.*

No one seemed to hear. There were happy giggles to my left. The watch became quiescent. And then the tremors of Leon's earthquake shook Nanga Sinyut to its foundations all over again. I got up and crept out for a swim.

The dawn was not fully visible. It was the time of day the Iban call Empliau bebungi, 'the calling of the gibbons', well after Dini ari dalam, 'dawn deep down', and just before Tampak tanah, 'to see the ground'. I made my way down the indistinct paths towards the river.

A 'swim' in Borneo is always a euphemism for something else: a wash, or worse. Taking the upstream track I soon came to the well-worn steps, cut into the mud bank, which led down to the river. Wading in, as the hanging mist over the river began to lighten and lift, I knew I was in the right place, the men's bathing area. I knew because it was a superior position, about fifty yards upriver from the women's area, where they swam, and where they collected the tribe's water; and I knew because the catfish swarmed round me, brushing my thighs with their long whiskers, nuzzling my y-fronts with their soft snouts, wanting me to take them down, hungry for their breakfast. A bit unnerved by their intimate attention I swam a little way off, into an upstream eddy by the bank, and there, luckily, all alone, I learned by personal experience the most important lesson of all for the tranquil conduct of life in the jungle: never, ever, shit in a whirlpool. It is a terrible decision to have to make, whether to duck or jump.

Sitting about on the river boulders to try and squeeze a little water out of my pants (bathing naked is not done in Borneo—the men go to great lengths not to expose themselves in public, and, besides, every little boy and girl is dying to know if your willy is as white as your nose and will hide in any available bush in an attempt to settle the point) I re-dressed in my dry clothes, washed my wet ones and left them on the rocks to dry, an automatic process which would begin at about 8 a.m. at Mansan jimbio, 'time to dry things in the sun'.

Feeling smug, lighter, and energetic, I set off to return to the longhouse, wondering secretly if James would perform in the river in front of everybody, like a man, or whether he would confer with himself and decide upon a policy of costiveness.

Musing on such deep concerns as the sky changed from orange to pale yellow to deep blue, I took the wrong path, the downstream way along the river bank rather than the men's route back to the longhouse. Hearing laughter and splashing and the shrieks of children I realised that I had unwittingly come upon the women's bathing place, by a little-used secondary path, and that I was still concealed from them by bamboo clumps, young palms and reed-like grasses. Feeling no more than a little ashamed of myself, I crept off the track and peered through the vegetation. An enormous tree had been thrown into the shallows and abandoned by some past flood, its great bulk forming a breakwater to the current and a safe lagoon in which to swim. It rose gently out of the water, and its branches were so many gnarled and tapering diving-boards for about twenty very young, very excited children, who swarmed up its trunk, ran out over the river, jumped in, swam to the bank, and then repeated the process, yelling all the while. Their mothers, slim and supple and half-naked and almost equally at home in the pool, were washing themselves and their sarongs, diving to wet their long black hair, collecting water in gourds, or in long segments of bamboo to carry home in baskets on their backs, or, a little way downstream, squatting near the river bank. Further out, a Brahminy kite, obviously a bird it was taboo to blowpipe, was quartering the river, gliding and flapping on its broad brown wings, fishing.

I returned to the proper path and walked back to the longhouse, past friendly groups of Kenyah men on the way to their stretch of water. Everyone was awake. Plumes of smoke rose in a regular row from the fireplaces at the back of the longhouse, straight up into the sky.

In the chief's room, the breakfast rice was boiling in iron cauldrons; and, just for us, a greasy stew of monitor lizard was exhaling its warm and stale breath from a Chinese cooking pot. It was a bit like the smell a dentist releases when he opens up an abscess to drain it. Feeling queasy, I was suddenly attacked by stomach cramps. Perhaps the feared moment had come at last? Perhaps the Borneo diarrhoea that gives you two minutes crawling time between the onset of heavyweight boxing in the stomach and a punch through the anal sphincter had finally searched me out? I got to my Bergen, found the right plastic bag with shaky hands, and, with a swig of cool chlorine water from the SAS bottle, swallowed two Strepto-triad and three codeine phosphate pills.

I sat very still, and, gradually, the pain began to subside.

'Makai?' asked Leon, sitting down cross-legged to rice and monitor with Dana and Inghai and a silent, hung-over James.

I shook my head.

'You the running shits,' said Leon, without a moment's hesitation, thumping his stomach to show me where it hurt. 'Me too, very often.' And thereupon he began to eat square yards of rice, acres of haunch of lizard; Don Giovanni must have been a horrible sight at breakfast, I thought.

Leaning back against one of the ironwood uprights, I had time to look slowly at the contents of the room. These people were Iron Age farmers. They were as primitive in their methods, perhaps, as the inhabitants of Skara Brae or the builders of Stonehenge. But, amongst the spears and blowpipes, the head-dresses, the leopard-skin coat, the decorated baskets, the closely woven padi-bins, the mats and tapestries, the wooden tools and crude axes, there were some very odd trophies indeed, prized relics of trading expeditions. There were eight tins of Brasso neatly arranged in a corner. A picture of the last Pope but one was impaled on a wall with four fish-hooks. A dusty cassette-player sat on the floor with two tapes of pop music beside it; a sewing machine, gold letters on its barrel proclaiming it to be *The Standard,* sat on show on a table; and behind it, most intriguing of all, there was a globe.

The pigs snortled and squealed under the floor; dogs, being gathered for a hunting expedition, yelped; the Iban went off to inspect the boat and debate the river level; James took out his notebook and *Les Misérables;* the rhythmic, soothing noise of the women pounding padi on the verandah began; the boxers in my stomach, still hitting each other feebly, were obviously in a clinch, and tiring; and I fell asleep myself.

Some hours or minutes later, I was shaken awake. An old man clutched me by the arm. He pointed at the medicine bag and tugged at my shirt. He was obviously much distressed and I followed him out into the gallery, carrying the first aid kit. Along the great raised passageway women were pounding padi by the bilek [family room] doors. Standing in pairs with heavy poles as tall as themselves, they alternately struck the rice, held in an ironwood trough at their feet. Their backs straightened and dipped, their hair fell over their shoulders, their breasts rose and flattened on the uplift, and fell, forward and full, on the downward stroke. Leon's girl, his moon in the sky, extraordinarily beautiful, smiled shyly at me.

To my surprise the old man led me right to the far end of the longhouse, past another group of older women who were weaving mats from strips of split rattan and two who were weaving cloth on six-foot-long wooden frames. We clambered down the notched log and set off along a path shaded by the huge leaves of planted banana trees. Rounding a corner, we came to a group of huts, all built on stilts, like a longhouse in single sections. The old man climbed the notched pole into the first one and beckoned to me to do likewise.

It was dark inside; and the stench seemed to soak into me. A circle of people, presumably the old man's family, stood round an old woman, presumably his wife. She was sitting on a stool, a bundle of tied sticks in her hand, fanning her foot. As my eyes adjusted, I looked where everyone was looking: at her foot. My stomach turned again. The top surface was an open pool of fluid with a clearly-defined, raised shoreline of indented flesh. She moved slightly as she fanned herself and, as she did so, yellow and black and red islets of infection slithered gently to new positions on the watery surface of the wound. The sons and daughters looked at me, enquiringly. An earnest young man mimed someone entering the river and treading on—a fish. She had stepped on a fish-spine. Her sarong was pulled up to her waist and her leg was a dark reddish-brown right up to her thigh, about six inches above the knee.

She looked at me, her face resigned and dignified despite the pain, but her eyes big and brown and pleading. It was a terrible moment. She had, I supposed, gangrene. And she needed massive doses of penicillin, far more than we possessed. I gave her two tubes of Savlon, two packets of multivitamins, and a roll of bandages. In return, the old man gave me three sweet potatoes, which I took. It was the nastiest transaction of my life.

Distractedly, I walked off into the secondary jungle, the near-by land on its fifteen-year rest between crops, to be alone. A Spiderhunter called somewhere; and I thought I saw an Orange-bellied flower-pecker, a small burst of flame dancing from bush to bush. I sat down, halfway up the hill, in sight of the longhouse, beside some kind of large, purple flowering orchid. Ought we to forget the Tiban range and take the woman to hospital in Kapit, a river's-length away? What ought we to do? So this was how the people of the far interior died—exactly as we had been told—of septicaemia, of one misjudged cut with a parang in a clearing, of a scratched mosquito-bite that became a boil, of a fish-spine in the foot. No wonder the population was so perpetually young, so beautiful. Perhaps Lubbock had got it the wrong way round in his *Prehistoric Times*. Perhaps it was not so much the 'horrible dread of unknown evil' which 'hangs like a cloud over savage life, and embitters every pleasure' but the very sensible dread, in this climate, of every passing accident, of every present micro-organism. They were certainly very stable societies, but perhaps this was exactly why they were so stable. The Niah caves excavations have revealed that the Borneo peoples of the true Stone Age ate the same kind of animals and made the same kind of boats as their probable descendants do today; and that the large mother-of-pearl fastenings on the ceremonial belts of Dana's daughters in the longhouse at Kapit are exactly the same as those worn by their ancestors forty thousand years ago.

When I got back to the longhouse James and the Iban were eating a lunch of rice and fish. I gave them the sweet potatoes; and I shared the problem with them.

'She dying,' said Leon, 'everyone know.'

'But what would happen if we weren't here? Could she get to hospital? Ought we to take her downriver?'

'She very poor. Very poor. All those peoples there just come from Mahakam. They not yet in the longhouse. She have no money for petrols to come back from hospital. She not want to go. She never left her families in all her life; and hospital very, very far. Very difficult. She die soons.'

'But what about the flying doctors?' said James. 'What about a helicopter?'

'Too long, very too far for helicopter. These people never see a helicopter. Too far for flying.'

Dana shrugged his shoulders and rolled himself a cheroot with Kenyah tobacco and a banana leaf. He spoke to Leon.

'The Tuai Rumah [Longhouse Headman] says we can take her to hospital on the way backs, if you want.'

The Iban had helped us to evade the issue, to cease being troubled. We sat and smoked in silence. Dana laid aside his native cone of strong king-size Kenyah and helped himself to James's Gold Leaf.

'The Tuai Rumah says the river too low to travel,' said Leon, 'but we make the water rise tonight. He know the customs. He very powerful man.'

Dana smiled his great chiefly smile, and waggled his finger at us, conspiratorially.

'We not to tell the Kenyah peoples because they harvest the padi. If they know we make our magics, they make their magics. They not want the rain.'

Inghai grinned. It was an Iban plot.

'They not have guns,' said Leon. 'They want to take our guns for tomorrow. Poof poof to shoot the deer. We say no, absolute noes—you hurt yourselves. They angry with the Tuai Rumah, just a littles.'

'Hang on,' said James, looking worried and making a characteristically emphatic gesture—both arms out, his palms up and horizontal. 'We do *not* want to upset these people. They're our hosts.'

'I know, I know,' said Leon, 'but the Rumah will raise the river and then we leave and then they not our hosts.'

A Kenyah woman came into the room, holding a necklace. She sat down beside us, and proffered the necklace to James.

'She want to sell it,' said Leon.

'What does she want the money for?' asked James.

'To send downstream with her husband, after the harvest, when the mens go to trade with the Chinese peoples.'

We inspected it in turn. It occurred to me that of all the presents we should have brought for the inland peoples, better than the aginomoto (the monosodium glutamate they all asked for), the sarongs, the salt, the cartridges, the batteries for their unused cassette player, the parang-blades from Kapit market, the ones they would have valued the most were the very objects which had seemed to us to be so obvious a nineteenth-century offering as to verge on the insulting: beads.

Beads, like the small brass gongs and the Chinese storage jars as big as a squatting-man (some of which date back to the Ming dynasty) to be found in almost every longhouse, are still a currency in Borneo. A few, made from shell and agate, were produced in the island; but most were imported by Arab and Chinese traders, some probably of Chinese manufacture, others from the Far East, and a few, almost certainly, from Venice. The most valuable of them all, the lukut sekala, a round black bead with delicate white and orange markings, used to be worth one healthy adult male slave, and would now cost well over a thousand pounds.

The bulk of this necklace was composed of yellow beads, labang, each one of which had been ground flat on the two surfaces that adjoined its neighbours by being fixed in the cut end of a piece of sugar-cane and rubbed against a smooth stone, probably about a hundred and fifty years ago. There were similar blue beads and seven much larger barrel-beads, black lukut, crudely painted with white, red and green stripes. The centre-piece was a pair of imitation pig-tusks, bound together with cloth, and fashioned from aluminium.

'Hey Redmond, I really want this,' said James, running the necklace through his fingers.

'We get it for you, my very good friend,' said Leon. 'Our Tuai Rumah will talk to their second chief and he will talk to the woman and you will get the necklace. We bargains.'

Dana and the woman went off to negotiate. James sat with his book, awaiting a conclusion of a sale; Inghai crept into a corner, curled up, and went to sleep.

'Come on Redmon,' said Leon, 'now we find place for magics.'

We climbed down the notched pole, made our way past the padi stores, and out along the river bank.

'We need a space to spin a pot,' said Leon.

'Spin a pot?'

'Yes Redmon. We show you. Our Tuai Rumah—he knows all the customs.'

'And you know a custom or two yourself, don't you Leon?'

'Sorries?'

'That girl in the pink sarong. She came to see you last night.'

Leon grinned and took his Homberg hat off.

'You very dirty mans,' he said, delighted.

'Leon, do you have a palang [penis pin]?'

'Who told you that word? How you know that word?' said Leon, genuinely startled.

'I read about it. I found it in a book.'

'Ah,' said Leon, 'you and Jams, you not ordinary mens. Jams always reading books.'

'He reads about palangs,' I said, grinning myself.

'Huh!' said Leon, pointing a muscly finger in my face.

'There's no need for you to smiles at the Iban. We know what you use.'

'What do you mean?' I said, standing still with surprise.

Leon paused and looked about us theatrically, checking the undergrowth for spies, glancing up lest there be eaves-droppers in the coconut palm.

'You,' said Leon, with great emphasis, 'use goat's eye-lashes.'

'Good lord!' I said.

'I thought so!' said Leon, mightily pleased.

'But Leon, when do you have it done? When do you have the hole bored through your dick?'

'When you twenty-five. When you no good any more. When you too old. When your wife she feds up with you. Then you go down to the river very early in the mornings and you sit in it until your spear is smalls. The tattoo man he comes and pushes a nail through your spear, round and round. And then you put a pin there, a pin from the out-board-motor. Sometimes you get a big spots, very painfuls, a boil. And then you die.'

'Jesus!'

· 'My best friend—you must be very careful. You must go down to the river and sit in it once a month until your spear so cold you can't feel it; and then you loosen the pin and push it in and out; or it will stick in your spear and you never move it and it makes a pebble with your water and you die.'

'But Leon,' I said, holding my knees together and holding my shock with my right hand, 'do you have one?'

'I far too young!' said Leon, much annoyed; and then, grinning his broad Iban grin as a thought discharged itself: 'But you need one Redmon! And Jams—he so old and seri-ous; he need two!'

Leon bounded away down the path, roaring with laughter and scattering a stray flock of hens into the bushes.

I caught up with him beyond the banana plantation. He was standing on the beach of a steep-sided small lagoon. 'Here we make our magics,' announced Leon. 'Now we tell the Tuai Rumah.'

We collected Dana, James and Inghai from the longhouse and took them to the witching grounds. Inghai carried a basket. James looked even more chiefly than usual, his new charm of beads and metal pig tusks laurelled round his neck.

I thought, as I fingered my own secret fetish in my pocket, a silver ankh given to me by my wife and all the other beautiful girls in her dress-making company in Oxford, of Edward B. Tylor's 1871 warning in *Primitive Culture:* 'In our own time, West Africa is still a world of fetishes ... Thus the one-sided logic of the barbarian, making the most of all that fits and glossing over all that fails, has shaped a universal fetish-philosophy of the events of life. So strong is the pervading influence, that the European in Africa is apt to catch it from the negro ... Thus even yet some traveller, watching a white companion asleep, may catch a glimpse of some claw or bone or such-like sorcerer's trash secretly fastened round his neck.' Still, if there was the very slightest chance that it would help us reach the headwaters of the Baleh and climb Mount Tiban with those terrible Bergens on our backs I was prepared to secrete anything that was not a worm or a bacillus or a virus or a python. In fact I would have settled for a totem pole, say, between my legs, or even a crucifix round my neck.

Inghai set down his basket and drew back the cover with his left hand, wiping his nose very slowly with the fingers of his right hand as he did so, a habit he indulged whenever he felt serious. The basket disclosed a wok, a packet of salt and a tortoise. Dana took out the wok and the packet and placed them on the shingle; he then leaped up the bank, drew his parang and cut down the branch of a palm tree.

Carrying his palm like a bishop's crosier, Dana indicated that we should do as he did. In a line parallel to the river we advanced to the wet stones, executed an about-turn and then marched ten paces back up the beach. Dana took a pinch of salt from the packet and buried it beneath a rock. We advanced back to the edge of the swirling river. Dana raised his palm branch, began a rhythmic Iban chant, and beat the water in time with his song. A much-startled Black-capped kingfisher, hawking insects from a perch on a dead tree to our right, dropped the dragonfly he was carrying and streaked off upriver, bright red and blue and yellow and disappearing fast.

'Good lucks!' said Leon. 'He go up; we go up.'

Dana waded into the water, chanting softly as he went, beating the river before him. Bending down, he planted the palm branch between the stones of the river bed, leaving about four feet of frond above the surface, where it fanned over, bent and trembling in the current.

'In the mornings,' said Leon, 'the river cover the palm.'

'Our Tuai Rumah,' said Inghai, deeply moved and bursting into speech, 'we very lucks. He know the adat lama [the Iban laws and customs]. He great man.'

Dana shouted to Inghai who then ran up the shingle, collected the wok and the comatose tortoise, took them out to the Lord of the House and returned to take up his position in the line.

Dana placed the tortoise in the wok and the wok on the concentric rings of a small eddy, mumbling very fast to himself as he did so and, with both hands, spinning wok and tortoise seven times to his left and seven times to his right.

'Hang on,' said Leon with great earnestness, in a turn of speech caught from James, 'now you say your England magics.'

'Inglang magics! Jams! Inglang magics!' danced Inghai, much excited.

James paused and then rolled his enormous eyes backwards into his bald head until only the whites were visible. It was horrible to look upon. Leon and Inghai studied their toes on the pebbles, uneasily.

'On your feet, young river,' intoned James, readjusting his corneas. 'We goad you on the bum with palms. We tempt you with salt, like the salt in the sea. We raise you up to the air, as a tortoise breaks to the surface.'

'Steady on, James,' I said, unnerved, 'we'll have an eighty-foot flood on our hands.'

We hurried back to the longhouse. Dana, in addition to all his other abilities, was an accomplished meteorologist. Huge clouds were beginning to stack themselves in the east.

We finished the monitor lizard and some sebarau that Inghai had smoked, and then began to pack our Bergens for the journey in the morning. The chief's younger sons wandered in and out, themselves preparing for the trip to their outlying padi fields on the following day.

I sorted the disarranged contents of all my watertight plastic bags, took my malaria pills, counted the morphine syrettes just in case Dana or James had been injecting themselves in order to sleep through Leon's antics, and, when James was not looking, I practised my subcutaneous perforation technique on a lone banana. As I pushed down the plunger on my water-filled spare syringe, stuck in the fruit, a jet of liquid hissed out the other side and shot across the room. I resolved, in the event, to inject James in a buttock rather than an arm.

Sealing up the medical kit and placing it in the middle of the Bergen, well padded with socks, I sat down by the central tallow lamp as night came down, and began to look again with delighted disbelief at all the montane and submontane species which Smythies illustrates in *The Birds of Borneo*. The resident old woman, stopping her weaving of small pieces of fishing net, came and squatted down beside me on her haunches. I turned over the plates, very slowly. She bent forward, intrigued, and her distended, looped earlobes, weighted with some twenty brass rings apiece, cast two ellipses of shadow across the rough planks of the floor. It seemed to take her some time to realise that the pictures were images of birds, birds that she knew; and then she uncurled a thin arm from around herself and pointed with a creaky finger on which all the joints were swollen. It was Plate III, the Borneo raptors, and she pointed at the Brahminy kite, *Haliastur indus intermedius*. Tentatively, she stroked its red-brown back; and then she turned, her old eyes alight, and she smiled at me with one set of lips and one set of gums.

The Kenyah chief's two sons walked over to us and one or two men came in from the verandah. I continued to turn the pages. They nodded with recognition and pleasure and talked excitedly to each other; but there were obviously some matters of weighty dispute. James and Dana and Leon and Inghai joined us, and everyone sat down in a circle.

'Bejampong,' said Leon firmly, putting his finger on the Crested jay, *Platylophus galericulatus coronatus,* a brown perky bird with a white crescent at the back of its neck and a plume like a second tail growing out of the back of its head. 'He very cheeky, like Inghai. And he talks a lots, like me.'

'Wass that?' said Inghai, sleepily, giving Leon a push.

'Very important bird for we Iban,' said Leon. 'He sing like hot sticks. We must hear him, after we chop the trees but before we burn the hills, to plant the padi.'

Leon mimed the fire with his fingers flickering like flames and made a rapid, crackling cry. The Kenyah nodded.

'See,' said Leon, taking the book, 'they agrees with me. And you must hear the bejampong before you goes hunting or for fights. He very quicks; you be very quicks. And his jugu'—Leon pointed at the jay's crest—'is like the hair on the head of a man you don't like'—Leon held up a patch of his own hair with his free hand—'and so you will take heads.'

There was an awkward silence, and Leon, realising that he had got over-excited and spoken out of turn, sheepishly handed *The Birds of Borneo* to his Tuai Rumah. Headmaster Dana proceeded to turn the pages with an air of authority, lecturing James and me, in official tones, in Iban.

'He the Tuai Burong,' said Leon, 'he know what the birds tell to us. Very, very difficults for ordinary mens. He dream dreams for chiefs. Not like our very naughties dreams, absolute no. Singalang Burong invite him to his house in the sky, to meet the birds, his—how do you say?—the husbands of his daughters. They called keptupong, embuas, beragai, papau, bejampong, pangkas and nendak. They look after we Iban. They speak to us and our Tuai Burong he understand.'

Dana held the book open for all to see, his thumb on the *Diard's trogon*, a long-tailed thrush-like bird with a black chest and a scarlet stomach, a fairly common but rarely seen resident of primary jungle up to about 4,000 feet.

'Pau, pau, pau, pau, pau, pau,' sang Dana in an ascending scale.

'He make the sound,' said Leon. 'Very good lucks sound this bird. Beragai laki and Beragai indu, the man and the wife in a bush. You can't see them. They laughs. You have good hunting; and then you laughs, too.'

Dana rifled through the plates and found a pair of Banded kingfisher, the male banded blue and black, the female black and brown, primitive tree-living deep-jungle kingfishers who are never seen over water.

'Pi-pit, pi-pit, pi-pit,' sang Dana, in falsetto.

'Very bad lucks,' said Leon, unknowingly disagreeing with Freeman's anthropological opinion that the Banded kingfishers is baka orang mentas jako, like someone speaking kindly. 'Embuas laki and Embuas indu—you hear them, you turn back, or you harms. If they fly mimpin, from your rights to your left, you runs back, all the way.'

At that moment there was a distant crack of thunder; rain

began to fall on the roof, way up above the great dim rafters, up above the feeble shadows cast by the lamp.

'Badas!' said Dana, with an enormous grin, forgetting himself and flexing both his champion biceps.

Inghai beamed with pride, and than looked with awe at his hero, the Lord of the House, the Bringer of the Rains.

'Our Tuai Rumah, he the best chief in all Kapit,' said Leon.

'Clever old Rumah zoomer,' whispered James, 'but I think he's been systematically pinching my ciggies.'

Seeing that the Kenyah were about to leave us and the party break up, I quickly dug Lord Medway's *Mammals of Borneo* out of my Bergen and opened it at a photograph of *Didermocerus sumatrensis harrissoni,* the Borneo (Sumatran or Asiatic) two-horned rhinoceros.

'Leon, ask them if they've ever seen this.'

Leon touched his eyes and then pointed at the picture, a captive female from Sumatra wallowing in her private mud pool in the Botanic Garden (Kebun Raya), Bogor, Indonesia.

There was much shaking of heads.

'Everyone heard tell of it,' said Leon, 'even at Kapit; but no one ever seen it.'

So then I tried them on a rare bird, confined to Borneo, the Bald-headed woodshrike, *Pityriasis gymnocephala,* for which Ernst Mayr, at the Museum of Comparative Zoology at Harvard, had asked us to search. 'The birds are slow and heavy in movements,' writes Smythies, 'keep to the middle canopy, and are difficult to frighten'; it had crossed my mind that the best way to observe them might well be to leave James out in a clearing and wait for them to come down to mate.

The grey-backed, red-throated, big-billed and bald-headed bird sat on its perch on Plate XLV (just below another Borneo rarity, the Great tit) and looked lugubriously at the assembled company through its little black eye. There was no answering look of recognition. No one had seen it; no one had heard of it.

Disappointed, I put my books away, and everyone went to bed. As the rain beat steadily on the atap roof I stayed awake, reading my Notebook of Useful Hints for life in the jungle, in the Mount Batu Tiban section under the sub-heading *Leeches*. In Oxford I had abstracted a paper by Smythies in the *Sarawak Museum Journal* for July–December 1957, but the whole subject now seemed less academic. Quoting Harding and Moore's great work on *Hirudinea* Smythies remarks that 'Very little relating to the leeches of Borneo has been published and our knowledge of them remains meagre. Undoubtedly many species await discovery and description. They are to be sought as external parasites on fishes, batrachians, turtles, crocodiles, and aquatic birds, in the air passages of water-frequenting mammals, etc., and burrowing forms under logs and in the humus of the rain forests. The true land and tree leeches will present themselves without being sought . . . ' The Borneo land leeches belong to the family *Hirudinea*, 'the Ten-eyed blood-sucking leeches'; and a 'leech that appears linear when extended may be egg-shaped in complete contraction. When unfed and resting it may be greatly flattened, transparent, pale-coloured and rough, with protruded, alert, sensory pappillae, and when gorged with blood the same leech will be many times larger, distended, thick, opaque, dark-coloured and smooth.'

And one must be careful when drinking from rivulets in mountain jungle, because of the Thread leech which attacks the nose and mouth: 'The presence of this leech is usually advertised by bleeding from the nose, and various Dusuns reported having seen wild animals, such as rats and mousedeer, infested by small leeches ... They are local in distribution and apparently confined to clean mountain streams ... The danger of this leech lies in its being very inconspicuous and having an exceptional reach for its size; it stretches to a thread. The Dusuns almost invariably assemble a spout made of a cleft of bamboo or of leaves when they require to drink from streams . . . '

But all that was as nothing when compared with the Giant leech:

'John Whitehead (*The Exploration of Mt Kinabalu*, p. 165) writes: "Tungal brought in an enormous leech; when it reared itself up it was quite a foot long of a pale cream colour. He found this horrible creature in the pathway close to the camp ... " Audy and Harrison (1952: 2) also found leeches of this type on Kinabalu: "Two distinctive and very large leeches were also encountered. One giant pallid leech came from the bridle-path near Tinompok. The other, darker and smaller, came from Kambarongah at 7,500 feet"' And, most disturbing of all: 'Dr. Nieuwenhuis (SMJ 1956, VII,

7: 123) mentions native reports of giant leeches on Mt Batu Tiban.'

Should I tell James? Or would it be kinder simply to allow him to pull them out of his pants when the need arose?

Whilst debating this difficult matter with myself, I became aware of a small figure at the far end of the room. It was Leon's girl in the pink sarong. Her step was delicate, her body sinuous, her movements quick and light. She hesitated when she saw my torch, so I switched it off and lay down. She reached Leon safely, without discovery, and soon their happy giggles and whispers mingled with the now-gentle sound of rain on the roof.

I fell asleep and dreamed of Leon and his beautiful young girl; of palangs and two foot leeches; and of Harrison's leeches who 'when they sense the presence of a victim ... stand up stiffly on the hinder sucker with the straight rigid body at an angle to the vertical'.

Into the Heart of Borneo, The Salamander Press, Edinburgh, 1984, pp. 75–89.

18
The Iban Hornbill Festival

ANDRO LINKLATER

Andro Linklater is a journalist and writer who was commissioned by Time-Life Books in the early 1980s to write a book on the Ibans of Sarawak for a special series entitled 'Peoples of the Wild'. Linklater travelled in Sarawak with three companions: a photographer, Tony Howarth, an 'Anglo-Australian' anthropologist, Dr Michael Heppell, and a local Iban anthropologist trained in Australia, Dr James Masing. Our writer was therefore supported by considerable expertise on Iban culture and society.

Unfortunately, Linklater's work never made it into the Time-Life series. The Ibans, in their outward appearance at least, were far too modern and not 'wild' enough, and, in any case, the series was ultimately discontinued because the Publishers decided it was unlikely to be commercially viable.

Linklater therefore used his experience to write an eminently readable and humorous travel book. For a relatively brief stay among the Ibans, Linklater demonstrates that he is a perceptive observer who was able to penetrate various aspects of Iban culture and psychology, no doubt assisted by his two anthropologist advisors. The relations and exchanges between the four travelling companions are often as amusing and intriguing as their relations with their hosts. As Linklater notes in his 'Acknowledgements', 'This account is written as comedy because no other vein was suitable to the intensity and débâcle of our expedition.'

The following extract, Chapter 6 of his book, relates part of the events of a major ritual, a gawai kenyalang, *or Hornbill Festival, held by Jingga, an Iban man, in the village of Langga, located in the Katibas basin, a tributary of the mighty Rejang River.*

THE curious thing about Jingga was that although his hair was grey and he walked with a weighty, aldermanic tread, he could not disguise the impression that a mischievous child lurked beneath the trappings of middle age. He was easily bored by the sort of grinding work that Mandau the augurer enjoyed. He was inclined to joke about matters which Langga the headman took seriously, and for a man in his fifties it was doubtful whether he should have enjoyed dancing as much as he obviously did.

It was just possible to infer from Mandau's remarks that the misfortunes on Jingga's farm might have been avoided with a little more effort. 'A good dream does not bring in the harvest,' he would say, quoting an Iban truism, and more than once he remarked upon the long hours he himself had invested in guarding his growing rice from pests—both insects and animals—and the rewards he had reaped thereby compared to his less fortunate neighbours.

Perhaps it was the feeling that Jingga should settle down and apply himself more seriously to the business of farming that gave rise to Langga's dream about the *gawai kenyalang*. The festival was the undertaking of a mature man, and although Jingga might not qualify initially, the preparations necessary for its success would surely induce a proper sense

284

of discipline, quite apart from the consequences which could be expected to flow from the hornbill's triumphant journey. If that was the motive, the scheme went off at half-cock. Not only were there the financial disasters beforehand but now, on the eve of the festival, it was becoming clear that Jingga's hopes of wealth were centred not on hard work but on the acquisition of a rich son-in-law.

As one fluent in Iban, conversant with longhouse custom, and capable of presenting him with a boatload of magnificent gifts for the feast, Michael made an admirable choice. His existing family in Australia was no impediment. What Jingga wanted was someone who would support both Aching and her son, Sulang, from her previous marriage.

'If you had an Iban wife as well as your wife in Australia,' he pointed out carefully, 'you would not have to live here, you could send her money while you were in Australia.'

It went without saying that when needed Jingga's son-in-law would also provide financial assistance for him and the rest of the family. As an inducement, he offered Michael the greatest honour which was his to bestow—the role of warrior in his *gawai kenyalang*.

'I refused when he first suggested it,' Michael explained later. 'I told him I wasn't feeling well, and that it would cause problems with the photographs, but he was clearly hurt. Then he began to make difficulties about us being there so I had no choice.'

Jingga's selection made the other warriors look serious. It was not a light matter they were engaged upon. They had all previously assisted at smaller festivals, building up their own spiritual resistance until they could withstand the forces at play in a hornbill festival. In addition, each of them carried protective charms whose power might harm someone without defences or experience. Had Michael been an Iban, they would have had no doubts that the dangers were too great, but it was just possible that a white spirit could look after himself. Doubtfully they accepted Jingga's decision.

At the tip of the hornbill effigy's beak, a thin bit of wood still joined the upper and lower mandibles. With the festival about to begin, the hornbill had to be able to eat the offerings which would be made to it, and so now the beak was sawn open. The first of many cocks and hens had its throat cut and its blood drained into a bowl of wine to make a cocktail

'Panchallong or Tenyalang, Wooden Image of a Hornbill', from Henry Ling Roth, *The Natives of Sarawak and British North Borneo*, Truslove and Hanson, London, 1896.

which Jingga carefully poured over the hornbill's beak. Then he picked up the effigy and with a rhythmic bass roar of 'Hoo-ha, Hoo-ha', he paraded it up the platform and down the gallery. Behind him came two attendants, dressed like him in loincloth and feathered hat. At every doorstep they stopped so that each family could pour a libation over the carved beak. Once the hornbill had been fed, it was only necessary to settle the wages of the *lemembang* or bards for the festival to begin.

At the heart of all the rituals which were to follow was the *timang* or sacred chant sung by the bards. In it they would describe the epic journey undertaken by two messengers who have been sent to take a message to Singalang Burong inviting him to the *gawai kenyalang*. When he receives the invitation, the god summons his son-in-law to help him kill the enemies of the celebrant; then carrying the heads they travel from heaven down to the longhouse, meeting many adventures along the way. At the longhouse the heads are split open and the seed which spills out is planted by Singalang Burong and his company before they return to heaven. It is epic poetry in the purest sense of the word. Its subject matter is the action of gods and heroes: to tell it in its complete form requires five days and nights of ceaseless chanting.

To commit such a work to memory is a remarkable feat, but in addition to the song appropriate to the hornbill festival, there are three others for different rites, and until he has learned all four, a bard cannot consider himself fully qualified. It is an arduous calling, which begins in childhood when the neophyte joins the chorus which accompanies the bard as he chants, repeating the words of the epic line by line. A brief period of instruction under the care of a master then precedes the years of study needed to commit the epics to memory. Understandably many content themselves with learning only two or three, and the less elaborate versions at that. Nevertheless, even at its simplest, the bard's is no small gift and he is worthy of his hire, especially since he alone cannot participate in the flow of *tuak* at a festival.

Quite what his hire should be took almost two hours of delicate negotiation at a crowded longhouse meeting. Since all the richer families were celebrating festivals which required ritual chants, there were to be four or five bards, each with a chorus of two assistants, and the rate for the hornbill festival would affect all the others. The first step was to find out what limitations there might be to their recitation. One bard explained that at a cockfight that afternoon he had helped to fasten on the spurs, and that this might give an inappropriate sharpness to his chanting: did anyone object? No one objected. Another pointed out that he did not chant the most elaborate version: was that all right? It was. Jingga then cleared his throat and asked the most senior bard how large a reward he wanted. The old man refused to commit himself, saying that he did not know what his colleagues might want.

'How much are you getting?' Jingga asked a bard who was to chant for a rice festival.

The man demurred. 'I'm doing it for my brother,' he said, 'I would be ashamed to ask for a reward.'

Another deflected the question by saying that he was chanting at his own festival.

'Then how much are you paying yourself?' Jingga shot back, and won himself a laugh from the crowd.

Eventually the senior bard was prodded into suggesting a figure. For a *gawai kenyalang* he thought a fee of $17 Malay would be appropriate. Jingga turned to other matters, touching particularly on the prevalence of pigs in the rice fields. In short, the bard's price was too high. After a good deal of allusive talk, they finally agreed on $15 instead, and everyone was satisfied.

The next morning Bulik took one of the little pigs out of its sack and stretched it out on the platform at the head of the ladder. Then up the ladder came the five warriors who were to defend Jingga's spirit in the days ahead. With their arrival, the great event which had been three long years in the making finally started.

The leader of the warriors was a lean, wrinkled old man, clad in a loincloth, a surcoat made from the fur of a honey bear, and a hat bedecked in hornbill feathers. He held the point of his spear a few inches above the sacrificial pig's throat, then thrust downwards. It was an ineffectual blow, and the agonised squealing did not bubble into silence until Bulik drew a knife across the animal's throat. Almost at once a cock was waved over the old warrior's head, and Bulik's beautiful wife, Orin, pressed a glass of rice wine upon him.

In some cultures it may be polite for a guest to conceal his opinion of the host's liquor, but not amongst the Iban. As though he had bitten on a lemon, the warrior's lips puckered, his cheeks went hollow, his eyes clenched tight shut, and his eyebrows scrunched up like caterpillars. Then he cast an anguished look at the line of girls behind Orin. Each was dressed like her in a short woven skirt, a salmon-pink blouse, and a bodice of clinking, silver coins, and each one held a jug and a glass of the same sour liquid.

Having endured that ordeal, the warriors made an offering, the first of some thirty occasions on which the ceremony was to be performed in the next few days. Before each man were arranged the by now familiar three rows of plates containing puffed rice, glutinous rice and rice cakes. Deliberately, row by row, they transferred the ingredients to a bowl, pressed seven eggs into the rice in a circle, decorated them with tobacco and betel leaf, and planted a tiny cup of *tuak* in the centre.

Crouching just beyond the offering Tony attempted to photograph the line of intense faces without including Michael's pale skin. He was working with absolute professionalism, and only someone who had heard his outburst could have guessed at the fury beneath his concentrated expression. In fact Michael's face was more than routinely pale; his pallor was deathly and covered with a glaze of sweat, and beneath his glasses his eyes had sunk into dark sockets. The excuse about not feeling well had not been false. Other than the main actors, the scene seemed to attract little attention. The crowd talked and laughed among themselves, and continued to hit the dogs as heartily as on any other occasion. Only when Bulik produced a cock for the senior warrior's prayer did the talk die away.

A growled bellow of '*Howa—Howa—Howa*' summoned the gods to hear a prayer which called unceasingly for increase—increase of rice, of land, of family, of wealth. Sometimes the prayer talked of the challenges which the family faced and sometimes of the deeds they had accomplished. Often the gods were reminded of the richness of the feast they were being offered, but sooner or later the same words always emerged, '*Bolleh padi, bolleh duit, bolleh ringgit*'—Give us rice, give us wealth, give us cash.'

With a swirl of the cocks, the prayer was spread across warriors and spectators like incense wafted from a censor, and the bird was passed to the next warrior who growled in his turn for the essentials of rice and wealth. Then the cock was sacrificed. Those who wished a special blessing were dabbed with its blood, and one of its feathers was planted in the offering bowl.

At the end when Bulik darted forward with a pint bowl of *tuak* for the celebrants to drink, Michael declined his share. During the ceremony he had been swaying from side to side and looked ready to faint. Now, having made the offering, he staggered away to collapse in Jingga's living-room. His forehead was burning hot, and he had a hard, dry cough. He was running a temperature of 105°, and must have found coherent thought difficult, but with iron determination he diagnosed himself as suffering from either malaria or some pulmonary infection, and ordered us to feed him anti-malaria tablets and antibiotics. His colleagues had their own diagnosis. As they had feared from the start, his spirit was being

crushed by the power of their charms. So while Tony and I
dosed him on pills, they dipped their charms of pebbles,
rings and petrified wood in oil and rubbed them on his lips
and belly in an attempt to revive his weakly spirit.

Throughout that night and the following day he remained
ill. From time to time he was fiercely alert, but for much of
the period he drifted vaguely in and out of consciousness.
Yet forty-eight hours later, either Western drugs or the warriors' charms or his own will-power produced an extraordinary change, for his temperature returned to normal and he
was on his feet again. Long before then the news had spread
up and down the Bangkit that the white spirit had succumbed to the power of the Iban spirits. When other details
of Jingga's hornbill festival have faded, that lesson will still be
remembered.

'I never really believed in the power of the spirits until
then,' a young man said later to James Masing. He had been
to school and the traditional beliefs of the longhouse had
begun to appear as superstitions. 'Now', he declared tri-
umphantly, 'I know that it is true—they do have power.'

While Michael lay sweating and half-conscious in Jingga's
apartment, the other warriors built a shrine round the central
pillar on his gallery from a frame of bamboo covered by a
ceremonial blanket. Inside it they placed the essential tools of
the rice farmer—the sacred seed and whetstones, a bush-
knife, and a trough for husking rice—and the three most
prized possessions of the family, a brass tobacco box, a gong
and a jar. Over the shrine they hung a gourd containing some
seeds of rice, nuts and fruit, and an empty grinning skull.
This, the Iban's ultimate fertility symbol, represented 'the
seed of the shrine', the principle of increase for which all
Jingga's men had prayed. And between the shrine and the
apartment door a length of red thread was tied which represented the obstacles which lay between him and their
prayers' fulfilment.

Parades now seemed to follow each other without cease.
Many of the activities on Jingga's stretch of the gallery were
repeated for the festivities which Mandau the augurer and
Langga the headman were celebrating, and to a lesser extent
by every other family. A cacophony of gongs and drums sig-
nalled the appearance of the girls and young women, brightly
dressed in beaded dresses, jingling coins and tall, glittering

head-dresses. They were led by a bard carrying his stave of office who danced along flicking his elbows and heels, and at their rear marched Jingga shouldering a flag and dragging a piglet in a basket. As they came back down the platform, the girls scattered rice over the sacrificial pigs to distract any evil spirits. Almost at once the warriors donned their finery and strapped long fighting swords to their waists. By a superhuman act of will, Michael staggered from his sickbed to join them, but this time his colleagues were taking no chances. They buckled a short bush-knife round him so that the spirits would not be further annoyed by his temerity.

The warriors' parade began bravely. The leader approached the shrine holding a naked sword in one hand, a hen and a bamboo leaf bouquet in the other. To symbolise how the obstacles to Jingga's success would be cut away, he slashed through the red thread between the shrine and the apartment, and to show how the difficulties would be swept away he whisked the bamboo leaves across the floor. Tense and half-crouched, he moved with a dancing, boxer's shuffle round the shrine then sprang at it, as though on an enemy, and with a whoop of triumph slashed the air inches from the basket. It was a dramatic performance, and the evil spirits threatening the increase which was about to come to Jingga fell away, cut to ribbons.

After the warriors had decimated the forces hostile to him, they removed their surcoats and swords and passed them on to the other male guests. These too showed a determined face in their slaughter of Jingga's enemies, but their successors were younger and more self-conscious. What had been high drama quickly deteriorated to the standard of amateur theatricals. While the old men had been serious, the young moved clumsily and the war-whoop reduced them to nervous giggles. Unable to bear the sight of Bulik shuffling round the shrine wearing bathing trunks and smoking a cigarette, I turned to watch the ceremony on Mandau's part of the gallery. The contrast was illuminating. Here even the youths circled, whooped and stabbed with savage concentration. Their loincloths were knotted more flamboyantly, their sword scabbards were more ornate, and their surcoats seemed more furry. A horrid suspicion about Jingga's surcoats now hardened to certainty. Mandau's men wore the skins of honey bears on their backs, as had the most senior of

Jingga's men. But the rest of his warriors—I looked again—yes, Jingga's warriors were clearly wearing deep-pile nylon rugs.

That was almost my last coherent thought, for offerings were being made in all parts of the gallery, and the *tuak* which followed flowed in spate. Without warning the bards began their incantations. In teams of three they strolled up and down the gallery, the leader striking the floor rhythmically with a long staff which had a clapper on the end. Their nasal drone, which was to continue for several days and nights, was at first strange, then irritating and finally hypnotic.

The fierce, familiar cry of '*Makai, makai*' woke me from a bardic trance, and I found that the gallery outside the central three apartments of Jingga, Mandau and Langga had become a dining-room. Twenty or thirty yards of floor were covered with plates of venison (I knew exactly where from), boiled hen, bamboo shoots, breadfruit, fish—boiled, broiled and baked—and rice in every form. It was the sort of banquet which in Europe went out of fashion in the seventeenth century, a potlatch feast indicating the host's readiness to break himself in order to satisfy his guests' appetites.

I ate with Langga's guests, and we did nobly I think, forging like ice-breakers through a sea of snow-white *tuak*. Our speeches were rich in prayers for the increase of Langga's wealth and rice, but when we sank back to the ground it was to find that the object of our prayers had repaid us by requiring us to drink not a bowl of *tuak* but a tumbler of brandy. Sometime later I can remember a girl singing me a praise-song, and her features behaved like a kaleidoscope. Her ears multiplied round the rim of her face, her mouth split into a dozen different orifices, and a necklace of eyes circled her nose which swelled and shrank with each note of her song.

The squeal of a pig being slaughtered awoke me. It came from the platform, an indignant cry which ended in a bubbling grunt. The sun had not yet appeared and a grey mist hung over the hilltops, but the platform had already been draped with blankets so that one section was divided off from another. There were five pigs on Jingga's section, and their deaths came quickly. A long-handled knife resembling a cut-throat razor was drawn across the throat, and as the pig gave its dying kick a deep incision opened the belly from ribs to groin. A twist of the body emptied out the hot entrails, from which the liver was deftly cut free to be put on a plate for divination.

Divining a pig's liver can be as complex as astrology, but its elements are simple. The left hemisphere pertains to the gods, and the right to man. Ideally the flesh is healthy and firm, and the bile duct well-inflated. This indicates a prosperous future for the individual whose fate is being divined. To improve their chances, the Iban only sacrifice young and apparently healthy pigs. The only badly diseased liver I saw was on the Batang Ai when a pig was killed for a distinguished head-hunter called Mani. It was blotched and shrunken, and Mani, as it happened, was terminally ill. The subtleties of divination lie between these two extremes, in deciding the meaning of minor blemishes and indentations of the flesh. A brief glance at the steaming organ suggested that Jingga and his family could look forward to the future with some confidence, but my hangover did not permit detailed examination.

A few yards away the bards circled round the pillars of the gallery and chanted of the search for Singalang Burong:

Have you seen the seed of the shrine decorated with tassels of human hair?
Have you seen the seed which is red like a blooming flower?

Some hours later the hornbill took flight. It was slotted into the top of the pole which had been prepared for it, a solid tree trunk, which was then carefully manoeuvred off the end of the platform and into a hole dug for it so that it stood like a flagpole beside the longhouse. Decorated with flowers and flags, the hornbill was assumed to go flying to the ends of the earth to find Jingga's wealth, and as though a load of responsibility had been removed, the festival then turned to comedy.

In a mirror image of the day before, more warriors appeared at the top of the ladder, but this time they were girls dressed quite grotesquely as men. In suits and shirts they strutted across the gallery where *tuak* was pressed upon them by simpering maidens with bass voices who had trouble keeping their sarongs knotted round their muscular chests. They paraded up the gallery, preceded by a matronly bard and followed by a shambling caricature of Jingga, hugely padded at the belly and dragging instead of his pig an empty bottle of arak.

Undisturbed by this display and the giggling crowd which followed, the bards sang of Singalang Burong's response to the invitations:

> 'I will not attend a feast in the land where red clouds form
> Until I have seen brains spattered all over the jar.'

As the day wore on the pigs' corpses on the platform swelled up to oval balloons in the heat, and their bloody mouths stretched to black, mocking grins. A boy, heavily made up with lipstick and eyeshadow, offered *tuak* to a girl who wore trousers and sported a charcoal moustache. A fighting cock tiptoed up to a sleeping dog and pecked it viciously on the tail so that it ran howling down the gallery.

Down on the beach where a fire glowed brightly in the gathering dusk, the nasal drone could still be heard from the longhouse:

> 'Then the slack old skin of the drums is replaced with that
> of a Kayan,
> Tautened with toggles made from the bones of his fingers.'

The lamps were lit inside the gallery and illuminated another feast of meat and another flood of *tuak*, but this time it was the women who offered up the prayers for increase. The next morning the bloodstains on the platform had turned dark brown, and dogs slept amidst a litter of rice and palm leaves. A small boy urinated on a dog, the dog urinated on a mat, and no one paid attention. A great weariness reigned. The gallery was strewn with sleeping bodies. Gobs of red betel and fragments of food littered the floor. Empty bottles of arak were rolled against the pillars, and buckets of *tuak* stood unattended by apartment doors. Through the debris the bards still wandered, and now the seeds in Singalang Burong's skulls were behaving like lovers:

> 'The last of the seeds to be planted are sitting, passing a
> lighted cigarette from hand to hand,
> The first of the seeds to be planted are lounging with
> infants in their laps.'

In desultory fashion people began to rouse themselves.

Ading started to mend a torn net. Two girls amused a child by banging on a gong. Mandau's wife swept up her section of the gallery. The offerings were taken down from their poles and from the loft. Still the bards paraded, but now the seed was flourishing and multiplying;

'The seed is seen to be people, sitting and talking widely together,
The seed is seen to be people, standing and reminiscing in ringing voices.'

It was time for Singalang Burong and his company to return to heaven.

Wild People, John Murray, London, 1990, pp. 99–110.

19
In the Rain Forest with Penans

ERIC HANSEN

Eric Hansen graduated from the University of California at Berkeley in 1971. Subsequently, he travelled very widely in the Middle East and elsewhere, and wrote articles for American newspapers as well as Geo *and* Australian Geographer. Stranger in the Forest *was his first book.*

We are accustomed to reading accounts of travel by water in Borneo. Here, Hansen tells us about a journey on foot through the interior rain forests of the island. Hansen was not an explorer in the tradition of Lumholtz nor a tourist like Boyle; his travels were undertaken for daring and adventure, in the company of forest hunter-gatherers. For once the dust-jacket blurb does not exaggerate—'a gripping adventure story', 'a rare and intimate look at a vanishing way of life'.

In 1982 Hansen set out to cross part of the interior of the island, travelling light, and following old trade routes. After unsuccessful initial attempts to cross the border between Sarawak and Kalimantan, he decided to work his way from Marudi in the Baram River basin, by river and overland to the Kelabit highlands and then on to Kalimantan. He managed

*to hire two Penan guides, John Bong and Tingang Na, to take
him by forest trails from the Tutoh River to Bario in the Kelabit
highlands. From there he went into Kalimantan, guided by
two other Penans, Bo 'Hok and Weng, who were from East
Kalimantan.*

*During several months Hansen covered about 1,500 miles
in Central Borneo, and arriving in the Kayan River basin in
Kalimantan and travelling downstream, he then decided to
turn back to Long Nawang and find a return route overland
to Sarawak. He did this and travelled a further 800 miles to
the north.*

*In the following extract, which comprises part of Chapter 8
of his book, we join Hansen in the primary rain forest of the
Upper Kayan basin on his way to Long Bia with his two Penan
companions. He had recently met Pa Lampung Padan, a traveller in the forest, who was journeying from his village of Long
Peliran in East Kalimantan across the border into Sarawak.*

DAYS after my meeting with Pa Lampung Padan, a
hunting party of five Penan men and their dogs surprised us during our midday meal. The men, armed
with six-foot-long spears, were hunting for wild pigs. They
appeared unexpectedly, but after a moment of tension they
sat down for an hour to 'share news'. One of the men told me
that three hundred miles down-river there were two white
missionaries living in the village of Long Bia. By coincidence,
Long Bia was where I was headed.

I wanted to buy more supplies in Long Bia and send a tele-
gram to my family to let them know that I was safe. No one
had heard from me since I left Kuching. More important, I
wanted to get out of the jungle. I had become physically
exhausted and was rapidly losing my patience. The sense of
magic and wonder that had sustained my spirits this far was
disappearing. In the jungle, time usually seemed to stand still.
There was no sense of urgency or destination. The journey
was an end in itself. But now, having crossed the last mountains and with the possibility of speaking English for the first
time in months, a destination became implanted in my mind.
I began thinking about arriving in Long Bia.

Long Bia became an obsession. I assumed that the missionaries were isolated from the outside world, that they
would be fluent in the local dialects, and that they would

be sympathetic towards the local people. I assumed many things, even that they would enjoy my company.

Three narrow trade routes cross the great dividing range between the Kelabit highlands and Kalimantan, but because Bo 'Hok and Weng, like all Penan, preferred the deep shadows of the forest, we meandered through a maze of game trails that had no beginning or end. I had been with Bo 'Hok and Weng for nearly two months. This was new country for them, and our route zigzagged through the forest in an aimless fashion as we traversed the mountain forests looking for food. Bo 'Hok and Weng wanted to explore this new landscape, and they laughed at my frustration about how little progress we made some days.

'*Dawai, dawai*' (slowly, slowly), they would say. They had a point. Why should they rush? There might be *gaharu* [sandalwood] or stands of sago nearby. I didn't know where I was and had finally learned to keep my suggestions to myself. Bo 'Hok and Weng were the pathfinders, so we continued to meander through the rain forest. During this phase of the trip I remained thoroughly disoriented. I knew we were headed in a generally southeasterly direction and stopped asking, 'How far?' or 'How many days?' The questions were meaningless.

'If you haven't been to this part of the forest before,' I asked Bo 'Hok, 'how do you know where we're going?'

'*Mal-cun-uk*' (we follow our feelings), came the reply.

He made it sound easy. It wasn't.

My anxiety about wanting to get 'somewhere else' was partially due to the fact that I knew too many 'other places' in the world. For Bo 'Hok and Weng there was no 'other place' apart from the jungle, and I grew to envy their sense of place, their contentment with where they were. When I became anxious, I would embark upon extraordinary journeys in my mind. When, for example, a steep, muddy trail became impossible because of the leeches, I might imagine myself on a pair of cross-country skis, gliding across expanses of unmarked snow, a picnic lunch and a bottle of wine in my pack. The sight of bee-larvae soup could send me around the world to the Empress Hotel in Victoria, British Colombia, for afternoon tea and scones with freshly whipped cream and thick strawberry jam. Outside, a light snow would be falling on the passing traffic.

Even during this relatively difficult period, there were some days in the rain forest that were effortless and full of new discoveries. We saw tree-climbing pigs and flying snakes and lizards, and one day Weng brought me a leaf in the palm of his hand. When I touched the leaf, it stood up and walked around looking for a place to hide. The leaf was actually a cleverly disguised insect that blended in perfectly with the leaf litter on the jungle floor.

Also, Weng told me the story of a diving ant that launches itself from the rim of Lowes pitcher plant (*Nepenthes lowii*) and plunges into the insect-eating reservoir of digestive fluid contained within the body of the plant. The diving ant rescues some of the insects by 'swimming' them to the edge of the reservoir like a miniature lifesaver. Then the ant eats the insect.

There were also times when nothing went right. One particularly nasty evening I lay sleepless alongside Bo 'Hok and Weng, shivering beneath a damp bedsheet. In the darkness I got up to pee and, without realizing it, walked over a column of fire ants (*Leptogenys processionalis*). I couldn't see the little monsters, but I knew straight away what they were. After swatting at my ankles and toes with my free hand, my fingers were covered with enraged biting ants. Then I made the mistake of switching hands, and a fire ant sunk its mandibles into the head of my penis. That got me moving. 'Ohhhhh, shit!' I yelled in a panic. The bite was like a wasp sting, and the pain shot up the sides of my body like an electrical shock and made my scalp tingle. Swatting and dancing about, I remembered an Australian remedy for neutralizing jellyfish stings. You simply pee on the affected area. I was desperate; I peed on my hands and feet to get rid of the ants, and that seemed to help. I brushed the survivors away and washed myself in a small stream. Much relieved, I lay down in the mist-sodden *pondok* [hut], but soon became aware of a strange vibration shaking the sleeping platform. A moment later I heard a sudden snort, like someone gasping for air, followed by a series of muffled cackling noises. I propped myself on one elbow and discovered Bo 'Hok and Weng weren't asleep at all. The bastards were convulsed with laughter at my performance.

I wasn't the only one who suffered that night. Bo 'Hok and Weng didn't have mosquito nets, so they nursed a smoky fire with damp wood to keep away sand flies and mosquitoes.

The blinding cloud of hot, choking smoke kept us hacking and sneezing until dawn.

To add to the miserable scene, my feet and ankles were battered and painfully swollen from the sharp river rocks. Lying on the sapling-pole platform, I could feel my feet throbbing from small infections that had erupted on my ankles.

Bo 'Hok had many talents, the least endearing of which was his ability to fart. He had been at it all night, and by morning it seemed as if his reserves were truly without measure. With each purposeful blast he seemed to be bugling the sleepy jungle to life. Under his onslaught the sounds of the night-hunting creatures and insects gradually yielded to the chorus of monkeys, followed by the lilting song of the yellow-crowned bulbul (*Pycnonotus zeylanicus*). Months later, an ornithologist at the Sarawak Museum would tell me that the nightingale really had a much better range, but as I sat up in the predawn chill with arms clasped around my legs, chin on my knees, and eyes still closed, I became even more convinced that there couldn't be a more lovely or peaceful sound in the world. I was being blissfully carried along by the bulbul's song when the jungle harmony was suddenly shattered by one last tremendous fart from Bo 'Hok, who then got up to start the morning fire.

We had definitely been reduced to basics, and I was beginning to wonder, after 127 days in the jungle, how much more I could take. I was becoming numbed and seriously disoriented by the never-ending search for food and a place to set up camp. I was also constantly aware of the closeness of the jungle. Distant views were no more than one hundred feet, and I spent long hours on the trail watching where I placed each foot while scanning my legs for leeches. The experience of jungle travel had become stupefying. But, worst of all, I was now becoming increasingly conscious of time. My feet were infected as a result of the wreck of the longboat, and it was painful to walk. I had been in the interior long enough; I wanted out. I wanted a rest from the jungle, to sink back into the comforts of my own culture, if only for a day. I convinced myself that all I really needed was to be able to catch a glimpse of my own people. They would help me remember who I was and where I had come from. I figured the missionaries in Long Bia would restore my fading spirits. I might be offered a bed with crisp, sun-dried cotton sheets, perhaps a pillow. My mind reeled at the thought of such luxuries.

For the next four days we cut our way cross-country towards the village of A Pau Peng. Below that village the Bahau River became navigable for long stretches between the rapids. Those days were wonderful. The jungle, in contrast to what we had so recently passed through, was spectacular. Two-hundred-foot hardwood trees were draped with orchids, ferns, and moss-covered vines. Bright orange bracket fungus grew in fan-shaped steps on dead logs, and the buttressed tree roots grabbed the steaming earth like giant webbed fingers. Occasionally we could catch glimpses between the trees of long, green valleys and sheer rock walls. The air was thick with brightly colored butterflies, and at one point we stopped to watch hundreds of large flying foxes settling in the mass of overhead branches. Flapping and wheeling with leathery wing beats, their chattering shrieks finally subsided as each creature found the right branch from which to hang. In the air they looked prehistoric. Bo 'Hok brought one down with a round of bird shot, and we ate it for lunch. It tasted awful. The meat was full of sinew and lead pellets, and the unpleasant smoky flavor lingered in my mouth for several hours.

Bo 'Hok and Weng were not river people, and once we left the river they began to take a renewed interest in the journey. We started talking about the plants again and would stop whenever we saw something interesting. Weng pulled up an aromatic root called *lung*. 'It's used to prevent ghosts from making babies cry,' he told me. Later that day we collected *akar korek,* the 'matches vine'. Once lit, the dried vine smoulders for days and is a convenient way to transport fire. To keep leeches from climbing up our shorts, we pounded the vine *akar sukilang* to a pulp and rubbed the resulting white mash on our legs. *Akar sukilang* could also be used for fishing. Like the *tuba* root the Iban use, *akar sukilang* can be pounded and mixed with water then poured into the river to stupefy the fish and make them float to the surface.

We decided to stop walking at noon one day, and Weng showed me how to catch birds with the sap of the *talun* tree. We smeared the sap on thin sticks 18 inches long and wedged them into parang cuts at the tops of fruit trees. Malay lorikeets (*Loriculus galgulus*) arriving at dawn to feed in the treetops became harmlessly glued to the sticky perches. We then climbed up the trees to collect the birds. They showed little fear and became tame very quickly. We made bamboo cages for the birds and carried them on top of our packs. The birds were friendly, affectionate little creatures, and we fed them cooked rice and bananas. The original idea had been to sell the birds downriver to the timber-camp workers for 1,000 rupiahs each ($1.60 U.S., a day's wage), but after collecting twenty we opened the cages and let the birds go. Unperturbed by their captivity, they flew back into the jungle. There was no clear reason why we did this. I guess I felt I had enjoyed traveling with the little birds and no longer had the urge to sell them.

We gauged time by the light and temperature as well as by animal and insect activity. A couple of hours before dusk every evening, swarms of sweat bees would cluster on our lower legs and start daubing the rivulets of perspiration with their little black proboscises. They didn't bite or sting, but they tickled to distraction. They also lacked even a basic knowledge of insect survival. They refused to fly away. In a single swat I could kill dozens, and those were immediately replaced by others. When the sweat bees arrived, it was time to start setting up camp. From earlier walks in Sarawak, I knew that at precisely 6.00 P.M. black cicadas would start their evening chorus to let us know that nightfall was near.

One evening as the bees swarmed over my legs, I pulled off my pack and felt the relief of a cool breeze blowing against my damp T-shirt. The swelling of my feet was getting worse, and I estimated I had another three or four days before it would be too painful to walk. The first thing we did was light a fire to keep the sand flies away. We had stopped at the site of a former longhouse. I knew this because along the riverbank there were fruiting pomelo trees that had out-lived the last traces of the village. Within the thick, pale green skins of these giant grapefruit were pinkish yellow segments of cool, sweet fruit. After walking all day they were irresist-ible. I ate one after another until the corners of my mouth stung from the zest of the peel. We sat on a flat-topped knoll peeling our pomelos and soon noticed the movement of fish in the stream. Weng finished eating, stripped off his shorts and T-shirt, and walked naked across the gravel riverbank with the gill net slung over his brown shoulders and buttocks. Keeping his genitals modestly covered with one hand, he stepped into the narrow stream and waded across to a partially submerged log. I could see him suck in his stomach as the cold water reached his waist. He attached one end of the net to the log and let the other end drift free. The line of evenly spaced white plastic floats attached to the end of the net drifted in a lazy, undulating motion before settling in a straight line with the gentle current.

At a familiar sound of leaves rustling, my ears flattened and the back of my neck tingled with excitement. I turned my head slowly and spotted the wild pig. Bo 'Hok had already seen it. We froze as the huge boar wandered out of the undergrowth one hundred feet from where we sat. The pig paused at the base of a thick stand of bamboo and suspiciously sniffed the air. Standing upwind, he was confused by our presence and stood there long enough for me to locate a shotgun shell in the bottom of my bag. Without taking his eyes off the pig, Bo 'Hok—who was closer to the shotgun—held out his hand for the cartridge.

Before the smoke of the shotgun had cleared, we were upon the animal with our parangs. The back of the neck and the hind legs were slashed deeply, and we stood back as the thrashing animal bled to death. No more than two minutes had passed from the moment of first sighting the pig until it was lying motionless at the edge of the clearing. After the blast of the shotgun, all sounds in the jungle ceased. There was only the trickling sound of the river and the smell of gunpowder and blood. Three-inch, curved, yellow tusks were visible from between the lips, and the thick, wiry bristles that stuck out from the snout continued to twitch for a few more minutes. It was a big pig. What we couldn't eat or smoke that night we would have to leave behind. The flesh would putrefy within a day. The instinct to kill and eat was growing on me. It was the natural response of the hunter and something I had never felt before. We forgot about fishing. Weng gathered up the net and rejoined us as we began the task of butchering the pig.

On a bed of wide leaves we cut back the skin with our small-bladed, long-handled knives. We carried these knives in a folded-bark sheath attached to the parang sheath. These smaller knives are called *anak* parang (child of the jungle knife) and are used for delicate work. The warm skin was spread out like a blanket. Setting aside favorite pieces such as the liver, tongue, and the twin muscles found inside the lower rib cage, we jointed, boned, cubed, then shoved the meat onto 18-inch wooden skewers. These would be hot-smoked over the fire during the night. We wrapped the brains in a leaf and steamed them on top of a potful of backbones. The local people would usually have rendered the skin and fatty tissue into cooking oil, but we didn't have a large enough pot to do the job. Locals rated pigs in terms of how many finger widths of fat they have in cross section on their backs. One finger width is not so good; four widths is excellent. For trading, the oil has a much greater value than the flesh because it's easy to preserve and transport and because there is a big demand for it in the villages. The pig we had just killed had three finger widths of fat.

We recovered four of the nine shotgun pellets from the carcass and put them away for later use. The spent plastic cartridge would be reloaded another ten times, at least, before being discarded. The firing caps would be poked out of the casing with a thin iron rod, hammered flat, and reprimed with a match head. Black powder is poured into the cartridge, followed by cloth wadding and the recycled lead shot. The cartridge is topped up with melted beeswax to make it waterproof.

I took half a side of ribs and forced it onto two sharpened green sticks that had been shoved into the loose root-filled soil and leaned slightly towards the fire. While the ribs were cooking, Weng and Bo 'Hok cleared a sleeping area. They also built a rack over the fire where we laid out the skewers of meat and lengths of firewood. The light was fading rapidly as I went to search for a species of wild ginger. Crushed in half a coconut shell with a smooth river rock, the ginger can be mixed with salt and chilli. We boiled rice, and when the meat had finished roasting, we cut it up and dipped each piece into the mixture. The smoky aroma and crackling succulence of the pig blended perfectly with the fiery ginger *sambol* [relish].

Squatting on my haunches, I could feel the fat running down my forearms and dripping from my elbows. I was totally absorbed in the meal when I heard a vaguely familiar droning sound growing louder and louder. I looked up through a gap in the jungle canopy and saw, for perhaps five

THE BEST OF BORNEO TRAVEL

seconds, a commercial airliner. It was streaking its way across the evening sky. As I watched, the light from the setting sun transformed the plane into a golden comet followed by a pure white vapor trail. It was a strange sight from where I crouched by the fire surrounded by the darkening forest and night sounds. I explained to Bo 'Hok and Weng that the airplane was bigger than a longhouse, but was unable to explain what kept the plane in the air.

The plane disappeared, and we continued eating, but the significance of the incident continued to grow in my mind. The passing plane had allowed me to see clearly how much I had adapted to the jungle over the previous four months. I imagined myself sitting in the plane, looking out over the rain forest for the first time. The uninterrupted vista of treetops swept to the horizon in all directions. Far below, at the edge of a small clearing, almost hidden from sight, I could see myself seated by a campfire with two brown-skinned jungle men, speaking an unknown language. Nearby lay the steaming carcass of a wild boar. From that perspective I suddenly realized what an astonishing situation I was in: submerged in a sea of giant trees, hunting wild animals, a bloodied parang at my waist—the stuff of pure fantasy six months earlier.

Apart from my feet, I was in excellent health. My stomach had flattened, and I had developed an acute sense of smell and hearing. I could distinguish small movements of animals on the jungle floor and imitate many of the natural sounds of insects and birds. More significantly, I had shed my Western concepts of time, comfort, and privacy. When I first entered the jungle and let go of my margins of safety to become vulnerable to a place I didn't understand, it was terrifying. I had slowly learned, however, to live with the fear and uncertainty. Also, I realized that the physical journey was not the great accomplishment. The value of the trip lay in everyday encounters, and the destination gradually became a by-product of the journey. Again I reminded myself that it was my ability to understand the local people and adapt to their way of life that had allowed me to get this far. I then realized, for the first time, that if I had come this far into the jungle, what could possibly stop me now from reaching the east coast of Kalimantan? Yes, I was definitely going to make it. I had spent six years thinking about this moment, but instead of the giddiness and surge of pride I would have anticipated, I felt serene and confident. Not that I wasn't thrilled. I could feel my face flush and tears come to my eyes, and I tried to hide these emotions. My daydream of crossing the Borneo rain forest was going to come true; that knowledge gave me an incredible sense of power and self-assurance.

That night the jungle was perfectly black, and a cool wind blew through the shelter. Bo 'Hok and Weng and I fell asleep joking about the condition of the tattered shoes I had attached to my feet that morning with adhesive tape. The nylon uppers were so rotten from the climate and constant damp they had completely detached from the soles. When I could, I planned to stitch the shoes back together with nylon fishing line.

We reached the Saban village of A Pau Peng the next afternoon. My toes were swollen together, and the tropical ulcers that had started as small sores now dripped lymph down my legs and ankles like clear candle wax. On each of the previous three mornings the pain had increased, and I felt so dreadful I couldn't talk during the first hour of walking. I was feverish, and each footstep made me feel nauseated. By midmorning my feet would lose all sensation, and I could begin to relax. A Pau Peng was located on the upper Bahau River, and this is where Bo 'Hok and Weng would head east for another week to reach their own land at the headwaters of the Malinau River. That night we went over our journey day by day, and I gave them shotgun shells for their wage. We went to bed early that night in an empty farm hut, and before dawn Bo 'Hok and Weng nudged me awake to say goodby. They returned to the jungle, and I never saw them again.

The next week was a blur of strangers and backwater communities. After Bo 'Hok and Weng the quality of guides plummeted. The two-hour trip downstream to Long Berini was thoroughly forgettable. I enlisted the services of a geriatric betel-nut addict and a young tough who could not walk without a crutch because of a childhood accident with a parang. As boatmen they were totally inept, constantly running aground and choosing the wrong way through the rapids. We agreed on a half-day journey so they could return upriver to A Pau Peng before nightfall, but after one hour in the longboat they pulled in at the first village and went in search of more betel nut. They then demanded lunch from the headman's wife, and the three of them conspired to convince me to give them money for the meal. This is not the custom, and when I pointed that out, they all agreed. The guides (I never learned their names) ate like gluttons, downed two giant glasses of *arak*, belched horribly, then flopped over for a sweaty afternoon siesta on a porch dotted with dog shit. An hour later they staggered to their feet and announced in a threatening manner that it was time for them to return to A Pau Peng and that they would like to get their day's wage, plus rental for the long-boat. We had a nasty argument, and I paid them for half a day. They accepted my gift of tobacco then jeered at my generosity, because they knew they were trying to cheat me. This introduction to the people of Kalimantan came as a shock, and I suddenly felt very alone.

The people of Long Berini were working in the gardens, and the village was deserted until dusk. I went for a walk to find someone interesting or sympathetic, but what I encountered was village life at its worst. Mange-ridden, copulating mongrels were stuck back to back at every turn, and the naked children were filthy. The buildings were architecturally boring, which came as a surprise because the simplest rice-storage huts in Sarawak and Kalimantan can be visually stimulating. Beautiful utilitarian joinery is possible with the parang and adz if there is some attention to detail and a sense of proportion, but here in Long Berini the low level of craftsmanship was immediately evident, giving the village a cheerless, temporary appearance. The afternoon walk gave me an idea of what to expect that evening. I don't remember the meal, only the snaggle-toothed, cross-eyed, dumb-staring, *arak*-breath faces that surrounded me until I thought I might pass out from the effort of being pleasant. The people of Long Berini spent the entire evening apologizing for not having white sugar or tinned food. They were poor in the worst possible way. They had lost all sense of pride and village spirit. There was no music or dancing or any sign of weaving or basketmaking that might suggest simple pleasures in daily work. Chinese antique dealers from Tanjung Selor had made a sweep through the village the year before, and there wasn't a single rice-wine jar, headhunting knife, or other artifact to remind the people of their heritage. The little sympathy I could muster for them disappeared as soon as we began discussing guides and a dugout to take me away the next morning. We bargained for hours and finally agreed that a man would take me downriver in a dugout next morning for two sticks of tobacco and two shotgun shells. I was glad to leave that wretched village, but the next one was no better. Similar evening scenes were repeated for the next five days as I inched my way down the Bahau River.

The communities on the upper Bahau River have few opportunities to trade because of the fickle nature of the river. The water level constantly changes, and the people are isolated in the villages. The communities are poor, and when they see an opportunity for gain, they grab it. I must have looked like a portable shopping mall with my shotgun shells, colored beads, salt, fabric, and medicine. I realized that my Indonesian had improved, because I could sit down with as many as a dozen villagers and patiently negotiate a fair deal for guides. With revolting hairless dogs lifting their legs on my pack and the prospect of yet another meal of tapioca leaves and rotten fish, I was often tempted to pay inflated prices just for the pleasure of saying goodby to my hosts. Unfortunately word travels from village to village very quickly, and once it's learned there's a newcomer willing to pay more than the going rate, the negotiations become even more prolonged and complex. Extra charges are thought up, and a price will not be settled until every possible avenue has been thoroughly covered. Until an agreement was reached I would have to stay in the village. No one was particularly concerned if I stayed or left because as long as I was in the village there was the likelihood I would part with some of my possessions—the more the better. There was no easy way out. Travel arrangements required a minimum of two or three tough bargaining sessions.

The young man who took me from the village of Long Uli to Long Peliran is memorable because he was the very worst guide I had in my entire journey. He arrived at my room in a cloud of cheap cologne late one morning dressed for the muddy slog in long, skin-tight, white pants and a T-shirt that read: 'American Disco Fever'. A pair of 1950s-style sunglasses straddled his black, bristled skull, and a compass hung from his neck. Dangling from his waist was a souvenir-quality parang suitable for opening letters and spreading marmalade on toast. I knew it was going to be a bad day. A homemade shotgun was slung over one shoulder, and in his back pocket sat a little snub-nosed 22-caliber six-shooter. He waved the pistol in my face and exclaimed with a well-practiced sneer, 'Can shoot! OK, *Tuan*?'

'Sure thing,' I replied, and off we went for the three-hour journey. We walked at a totally distracting pace. Quick sprints interspersed with long stops and exclamations (his) of '*Lelah!* (tired) ... '*Susah!* (difficult). Sweat beaded up on our foreheads and streamed down our faces, and we would hurry off again. For an hour we continued sprinting, stopping, sprinting, and every so often the young man in white would turn to me on the trail and brandish his gun menacingly.

'Can shoot! OK?' he repeated. On this cue I would nod per-functorily, and we would plow on. It was madness to carry on like this, and soon I was thoroughly pissed off.

'*Tuan* can walk strong!' he finally gasped.

'Fuck you,' I replied brightly.

He refused to speak Indonesian, and our conversation was limited to his ten-word English vocabulary. His repeated garbled phrases became incomprehensible, and I had to cut him off. I taught him to sing: 'Zippity doo dah, zippity aye. My, oh my, what a wonderful day. Plenty of sunshine head-ing my way. Zippity doo dah, zippity aye.'

He practiced his song for the next two hours, and I was left in peace. I provided encouragement at regular intervals to maintain his interest, and by the time we marched into Long Peliran he was sounding terrific.

Long Peliran, a Kenyah village situated on the left bank of the Bahau River near Sungai Lurah, was approximately thirty-five miles downstream from where I had left Bo 'Hok and Weng six days earlier. It was the home of the proud pos-sessor of the Singer sewing machine. Although I had no rea-son to suppose that in any other way the village would be more cheering than the others I had visited in the last week, I arrived at Pa Lampung Padan's house in Long Peliran to find the people there delightful. Enshrined in the common room of Pa Lampung Padan's house was the sewing machine from Marudi. His wife was full of life and obviously delighted with her husband's great journey. I stayed for two days; then Pa Lampung sent me downriver with a long-boat full of friends. 'How much for guides and the boat?' I ventured to ask. He laughed at such a ridiculous question then filled my arms with a stem of bananas and skewers of smoked deer meat.

Once I left Long Peliran, I began covering incredible distances—fifteen to twenty miles in one day. The river widened to as much as 100 feet, and the sun became a problem. After seven weeks in the shadows, my skin began to burn. Each day or two I joined a different group going downstream, and the longboats I traveled in kept getting larger, until we were using the standard 40-foot, eight-man dugout canoes. With a skilled crew these boats could get through the really big rapids that lay between the village of Pudjungan and the mouth of the Bahau River.

Fifty years before, Charles Hose, the British Resident of Sarawak, had written this about upriver travel in Borneo: 'The Kenyahs are experts in handling these boats, and unlike the other tribes, seldom have accidents. It is a very moving sight to see a boat slowly gliding downriver to the head of a fall, the men standing up and leisurely dipping the ends of their paddles in the water to keep the boat's head straight, and straining their necks to find the best spot to shoot the falls. Suddenly they drop down on the thwarts and paddle for all they are worth, the boat dashing into the foaming mass of waves between huge boulders. The roar is deafening, and the water splashes in on all sides. For a moment it seems as if it would be impossible to get through, but pace tells, and by wonderful handling the boat, often full of water, slips around into a less troubled part, where the crew obtain a short breathing space in which to bale out the craft and prepare for the next rapid.'

This description is all too true. For three days we traveled in this fashion through six named sets of rapids and numerous unnamed smaller ones. I must have bailed tons of water with my dinner plate. If there were any lingering resentments about the upper Bahau River communities, they were quickly erased by the excitement of the river. This was the way to travel. Giant 8-foot monitor lizards swam in the shallows, and monkeys and large herds of wild pigs wandered along the big gray river rocks and open riverbanks. The hunting was fantastic. In every shade of green, a dense jungle grew 150 to 200 feet high on both sides of the river.

I traded two sticks of chewing tobacco for a large conical sun hat and could then luxuriate in comfort under the intoxicating, languid tropical sun. It was a relief to be out of the perpetual shade and dampness of the rain forest. The sunlight made my skin tingle, and my clothes were dry. I didn't have to wear my shoes, and my feet appeared to be healing now that I wasn't walking.

At the junction of the Kayan River, we switched boats for the last time, and I climbed into *The Pioneer*, a 60-foot open-deck river trader powered by three 40-horsepower outboard motors. *The Pioneer* was on her way downstream to the trading post at Long Bia and would arrive the next day before noon. Whether or not Long Bia lived up to my expectations, my arrival there would mark the end of my first crossing of the island.

Behind me lay four months and one thousand five hundred miles of jungle travel. The last day on the river *The Pioneer*, with thirty assorted passengers on board, started off in a jungle mist before sunrise. An hour later the blue sky and sunshine broke through. Logging camps began to appear, and we arrived in Long Bia just before the heat of the day. Most of the passengers were still sleeping soundly as the bow eased into the muddy riverbank below the trading post.

I climbed up the riverbank with my woven rattan backpack. It contained a blackened cooking pot, a bamboo section of wild bee's honey, my mosquito net, and less than 50 of the original 250 shotgun shells. The rest of my goods, the tobacco and beads, had been extracted from me by the villagers on the Bahau River. I looked at my tattered clothing and realized how wild I must look, even in a small village such as Long Bia. I smoothed out my khaki shorts and straightened the knotted camouflaged shoelaces on my ridiculous footwear and, just as I had imagined it weeks earlier, walked up to the white Colonial-style missionary bungalow that overlooked the river. How I had waited for this moment! Before I reached the second of five carefully swept front steps, I was greeted from behind the screen door with a surprisingly loud, 'Well, what do you want?' spoken with a strong accent from the American South. The voice belonged to Ian, the Mission Aviation Fellowship (MAF) pilot. He wore mirrored aviator sunglasses and, without opening the door, stood with his hands on the hips of his ironed trousers waiting for my answers.

'What do I want?' I echoed weakly. 'Well, I just arrived and thought you might like to have a chat.' It was painfully obvious to both of us that nothing could have been further from his mind. After weeks of looking forward to this meeting, I was at a total loss as to what to say. I continued uncertainly, 'If you're not too busy, that is … well, you know, I could always come back later on, I guess.' This was followed by an uncomfortable silence as we looked at each other through the screen door. My vision of Long Bia and a friendly reunion with a fellow Westerner evaporated.

I was backing down the stairs when Ian's wife, Julie, appeared beside him and invited me in. After removing my shoes, I was immediately led to the bathroom to wash up and was confronted by a brand-new bar of Dove soap, a white porcelain washbasin, and a blue terrycloth handtowel with matching washcloth. I looked at the baby blue flush toilet and the shower with the blue plastic curtain and suddenly realized that I really didn't want to be there. I had an almost uncontrollable urge to climb out the window, to be back in the rain forest looking at walking leaves and catching Malay lorikeets with Bo 'Hok and Weng. I wanted to chase a wild pig through the jungle with hunting dogs and a spear. The last place in the world I wanted to be was in that bathroom washing my hands with a bar of Dove soap. I reconsidered and then washed my dreadful-looking feet in their sink.

As I washed my feet, I glanced in the mirror above the sink and was shocked at what I saw. I had a penetrating half-mad look to my eyes, a look of such unblinking intensity that immediately I realized why Ian hadn't been eager to invite me into his home. I stared into my own eyes for what seemed like five minutes then towelled my feet dry.

On the verandah, Ian and Julie were sitting beside a white wicker table. Their two infant children, Michael and Janie, were playing on the polished hardwood floor. I sat down and glanced at what was on the table. I stared dumbly, and perhaps too intensely for my host's comfort, at four perfectly shaped chocolate chip cookies neatly arranged in the center of a sparkling white dish. They looked like some sort of offering or monument to the memory of white rule. At first, I was merely hypnotized by this apparition, then outraged. I touched one. It was real, and still warm. I wanted to make the cookies disappear. They were intruding on my fantasy, my vision of a dank jungle world of buttressed tree roots and wild-animal sounds. The cookies also smelled delicious. They transported me back to my grandmother's kitchen and the strawberry-shaped ceramic cookie jar of my childhood. I was overwhelmed by conflicting emotions, and in the end I did the only thing possible. I took the cookies in my hands and ate them. Without a word I ate every last one. After months of living in the jungle, it would be obscene to describe what those cookies did to the inside of my mouth. The sensation of white sugar, melting chocolate morsels, and the scent of vanilla essence. Ooooh! ... the decadence ... the saliva ... the shame!

Ian and Julie were perched nervously on the edge of their seats staring at me, perhaps waiting for me to disappear. I looked at the Christian paintings, calendars, and framed scriptural quotes that covered the walls. I had been in the jungle too long to make any sense of them. All these symbols seemed trite and out of place. And their puppy. That sweet-smelling, freshly shampooed, little bundle of white fluff would have been torn to pieces in thirty seconds by the village hunting dogs.

I started to explain my route from Kuching, but soon realized that Ian and Julie weren't listening. I sensed that I had interrupted something important. I was correct. After a two-month wait, they had just received a package from the United States. They were anxious to make sure everything was there. I let my voice trail off, and they gravitated back to their cardboard treasure chest. Yes, it was all there: the instant pudding, the taco and pizza mixes, the Kraft Parmesan cheese in its distinctive green container, Jell-o, some Dinty Moore Beef Stew in family-sized tins, and a jar of Orville Redenbacher's Gourmet Popping Corn. What can one say about these things?

After their excitement had died down I mentioned that I was planning on going to Data Dian, a village in the mountains where I wanted to do some trading. I asked Ian about a possible lift in the mission plane. Without a moment's hesitation he replied, 'No, we can't do that!' Ian went on to say that there was a backlog of vital mission supplies that had to be airlifted to the highlands. 'There's just no way we can take someone like you, there's no room.'

Their lack of interest in my arrival was a shock, but I reasoned they were there to do a job and to help the village people. I couldn't expect special privileges. The people should come first. I also realized that it must have looked very self-indulgent of me to be wandering around the jungle without any specific purpose that they could see.

We were all uncomfortable, so I decided to leave. I collected my shoes at the door and walked away from the house speechless. As I left I heard the 'click' of the screen door as it was locked behind me. That must have been their way of saying goodby. From a distance I could hear the sound of a child being spanked. The puppy barked.

I wandered over to the small corrugated metal mission warehouse next to the landing strip. I was curious to see what things were essential to village life in the area. I expected medicines and tools, perhaps educational material. I stood with my hands and face pressed against the wire-mesh window, peering into the dim interior. When my eyes adjusted to the dark, I could see the vital mission supplies that Ian had referred to. I listed them in my journal. There were Eveready flashlight batteries, Roma and Ayam brand tea biscuits, Baru Baru laundry soap, Tancho lavender-scented hair pomade, souvenir-quality headhunting knives from the coast, Jackson Super Milk Toffee Sweets, Super Bubblegum, linoleum, white sugar, tins of instant chicken noodle soup, infant formulas from Switzerland, soft drinks, Bliss Peppermints ('sweetens the breath'), and wind-up plastic penguins from Hong Kong.

It was all neatly arranged in colorful cardboard boxes. Looking into the storeroom, I caught a glimpse of my own culture. The taste of chocolate lingered in my mouth, and I could feel my roots.

The owner of *The Pioneer* had given me the name of Nyonya Nam Sun, the sister-in-law of the local trader, Tokay Moumein. I couldn't think of anywhere else to stay that night, so I asked directions to her house. Word of my arrival in Long Bia traveled quickly, and by the time I found Nyonya Nam Sun's house, everything had been prepared. Nyonya Nam Sun had been expecting me. The room had been cleaned, and there were frangipani flowers on a low dressing table. The mattress, though twelve inches too short for me, was covered with a clean, flower-print sheet. Nyonya Nam

Sun brought me sweet black tea, and twenty minutes later she produced a delicious lunch of fried fish and rice with a bowl of bitter-melon soup. It was good to be speaking Indonesian again and to be eating with my fingers.

'Would 1,000 rupiahs ($1.60 U.S.) be all right for the room and two meals?' Nyonya Nam Sun asked me.

'One thousand rupiahs is fine,' I replied, 'but 1,500 rupiahs would be better.' I was so touched by her hospitality that I offered her this extra amount. Nyonya Nam Sun smiled then brought a second helping of lunch.

I left Long Bia the next morning intending to end my journey across the island at Tanjung Selor near the mouth of the Kayan River. In a dugout canoe with six paddlers, we headed into the main current. We glided past timber camps and areas of a clear-cut forest. Log rafts became more frequent, and the river was dark brown and muddy from the erosion caused by hill logging. Almost every longboat had an outboard motor, and the people wore Western clothes. It was upsetting to see these things. Also, I worried about the police in Tanjung Selor. If they detected me, which was highly likely, I would have a hard time explaining my expired passport and non-existent visa. The real jungle was far behind, and we were rapidly approaching the downriver commercial centers. The sores on my feet had erupted again, and I wanted to go to a hospital. The missionary family had repulsed me with their self-centered, insulated existence, so I sat in the dugout, despondent, waiting for my adventure to come to its disastrous end.

I had lost my momentum and confidence. I felt so miserable I didn't bother to paddle. I sat on the duckboards like a doughy lump, feeling sorry for myself because after so much time and effort the trip was going to end badly. Maybe the entire project had been a farce sustained by my foolish pride? I felt like a fool, but during the next couple of hours I started to recall many of the remarkable moments of my journey. Faced with my imminent arrival at one of the squalid coastal boomtowns, I became uncertain what to do. I had to make an important decision. I asked myself: could I possibly contemplate a return to the jungle? I was exhausted, depressed, and half-crippled. The jungle had just spat me out like a piece of old chewing gum. And what were my chances of making it through another eight hundred miles of jungle in my condition? I could be back in San Francisco within a week, and the temptation to go was great. Get on a plane, I tried to convince myself. Don't be a fool!

I'm not quite sure what prompted me, but early that afternoon, 137 days after entering the rain forest, I decided to turn back. I was within a day's journey of Tanjung Selor. I ignored the confused protests of my Kenyah companions, and fifty miles from the ocean we made a slow, wide arc in midstream and started back upriver. '*Orang gila!*' (crazy man), one of the men muttered.

I waited ten days at Long Bia for my feet to heal. Following Nyonya Nam Sun's suggestions, I washed them several times a day in salty water, and by keeping the weeping sores covered with bandages, the swelling gradually subsided and the ulcers began to close. I slept with my feet elevated and spent my days collecting trail information from visitors and piecing together a hand-drawn map that, with luck, would lead me back to Sarawak. One day I spent four hours stitching the patchwork remains of my shoes back together. The first crossing of the island had taught me how to travel. The return trip would test my limits. I planned to recross the island by a much more difficult and uncertain route.

Stranger in the Forest. On Foot across Borneo, Century Hutchinson, London, 1988, pp. 153–74.

Other Oxford Paperbacks for readers interested in South-East Asia, past and present